MCAT® Critical Analysis and Reasoning Skills (CARS)

2025–2026 Edition: An Annotated Passage Guide

Copyright © 2024
On behalf of UWorld, LLC
Dallas, TX
USA

All rights reserved.
Printed in English, in the United States of America.

Reproduction or translation of any part of this work beyond that permitted by Sections 107 and 108 of the United States Copyright Act without the permission of the copyright owner is unlawful.

The Medical College Admission Test (MCAT®) and the United States Medical Licensing Examination (USMLE®) are registered trademarks of the Association of American Medical Colleges (AAMC®). The AAMC® neither sponsors nor endorses this UWorld product.

Facebook® and Instagram® are registered trademarks of Facebook, Inc. which neither sponsors nor endorses this UWorld product.

X is an unregistered mark used by X Corp, which neither sponsors nor endorses this UWorld product.

Acknowledgments for the 2025–2026 Edition

Ensuring that the course materials in this book are accurate and up to date would not have been possible without the multifaceted contributions from our team of content experts, editors, illustrators, software developers, and other amazing support staff. UWorld's passion for education continues to be the driving force behind all our products, along with our focus on quality and dedication to student success.

How To Use This Booklet

In the following pages, you will find copies of the passages used in the UWorld MCAT Critical Analysis and Reasoning Skills (CARS) book. Each passage appears twice in this booklet: once with annotations focusing on the passage's structural features, and a second time with annotations focusing on its content. The annotations are divided into categories for organization and ease of reference as shown below.

Structural Annotations	
Paragraph Summary	A brief synopsis of each paragraph
Topic Shift	When a passage switches its subject of discussion
Addressing Opposing View	Raising and responding to potential counterpoints to passage claims
Connected Ideas	Information found in a later part of the passage that relates back to a previous idea

Content Annotations: Passage Observations	
Claim	A statement that seems notable or important for understanding the passage
Focus	A statement that helps to indicate the passage's main idea
Author's View	Information that suggests or reveals the author's opinion
Emphasis	A point that the author particularly stresses or draws extra attention to
Likely Question Topic	Information that could easily form the basis for a question
Outside Source	A reference to ideas other than the author's
Elaboration	Information that provides further details about a previous claim

Content Annotations: Potential Pitfalls	
Misreading	Cases where it would be easy to make a mistake about what is being said
Details	Information containing many nuances or complexities to keep track of
Missed Distinction	Cases where it would be easy to overlook a point the author makes, or to assume the author makes a point that they actually do not
Trap Answer	Claims that could make an incorrect answer to a question seem attractive

You can think of these annotations as representing the kind of mental notes that a CARS expert would make while reading. Accordingly, reviewing them in context will give you a sense of the types of things that are important to notice when analyzing a passage. As you continue to practice and become more experienced with CARS, you may find yourself naturally making similar (but less formal) mental notes.

In the main UWorld CARS book, you can read and analyze the passages as they would appear on the exam (without annotations). Then, in this booklet, you can review the annotated versions of each passage.

Table of Contents

PASSAGE A: THE KNIGHTS TEMPLAR
Structural Annotations 2
Content Annotations 4

PASSAGE B: JACKIE IN 500 WORDS
Structural Annotations 7
Content Annotations 9

PASSAGE C: PROBABILITY AND THE UNIVERSE
Structural Annotations 12
Content Annotations 14

PASSAGE D: AMERICAN LOCAL MOTIVES
Structural Annotations 17
Content Annotations 19

PASSAGE E: THE DIVINE SIGN OF SOCRATES
Structural Annotations 22
Content Annotations 24

PASSAGE F: THE VICTORIAN INTERNET
Structural Annotations 27
Content Annotations 29

PASSAGE G: FOOD COSTS AND DISEASE
Structural Annotations 32
Content Annotations 34

PASSAGE H: FOR WHOM THE BELL TOILS
Structural Annotations 37
Content Annotations 39

PASSAGE I: MEANING: READERS OR AUTHORS?
Structural Annotations 42
Content Annotations 44

PASSAGE J: THE INKBLOTS
Structural Annotations 47
Content Annotations 49

PASSAGE K: LENGTHENING THE SCHOOL DAY
Structural Annotations 52
Content Annotations 54

PASSAGE L: WHEN DEFENSE IS INDEFENSIBLE
Structural Annotations 57
Content Annotations 59

Passage A: The Knights Templar

This passage is associated with 5 questions analyzed in the UWorld MCAT CARS book.

- Subskill 1a. Main Idea or Purpose Question 1
- Subskill 2d. Connecting Claims With Evidence Question 1
- Subskill 2e. Determining Passage Perspectives Question 1
- Subskill 3d. External Scenario Support or Challenge Question 5
- Subskill 3f. Additional Conclusions From New Information Question 2

Passage A: The Knights Templar

Structural Annotations

Passage Structure: 6 Annotations

[P1] The seal of the Knights Templar depicts two knights astride a single horse, a visual testament of the order's poverty at its inception in 1119. Nevertheless, these Knights of the Temple—who swore oaths not only of poverty but of chastity, loyalty, and bravery—would eventually become one of the wealthiest and most powerful organizations in the medieval world. So far-reaching was their strength and acclaim that their destruction must have seemed as sudden and surprising as it was utter and irrevocable. The signs of danger could not have been wholly invisible, however, as the Templars' growing influence became perceived as a threat to European rulers.

Paragraph 1 Summary

The Templars rose from humble beginnings to become extremely wealthy and powerful. Nevertheless, they came to a sudden and surprising end due to conflict with European rulers.

[P2] The first Templars were nine knights who took an oath to defend the Holy Land and any pilgrims who journeyed there after the First Crusade. Having secured a small benefice from Jerusalem's King Baldwin II, the knights inaugurated their mission at the site of the great Temple of Solomon. They quickly attracted widespread admiration as well as many recruits from crusaders and other knights. Within a year of its founding, the order received a financial endowment from the deeply impressed Count of Anjou, whose example was soon followed by other nobles and monarchs. As early as 1128, the Templars even gained official papal recognition, and their wealth, holdings, and numbers swelled both in the Holy Land and throughout Europe, especially in France and England.

Paragraph 2 Summary

Because of the great admiration they inspired, the Templars rapidly gained members, wealth, and influence.

[P3] [1] However, this growing power contained the seeds of the order's downfall. Although the Templars were generally held in high esteem, the passage of time saw censure and suspicion directed toward them. The failure of the disastrous siege of Ascalon in 1153 was attributed by some to Templar greed. Similarly, in 1208 Pope Innocent III condemned the wickedness he believed to exist within their ranks. Moreover, their increasingly elevated status brought them into conflict with established authorities. One revealing example occurred in 1252 when, because of the Templars' "many liberties" as well as their "pride and haughtiness," England's King Henry III proposed to curb the order's strength by reclaiming some of its

[1] **Connected Ideas**

The author revisits the idea from the introduction that the Templars' power would eventually lead to their downfall. This relationship is likely a major theme of the passage.

possessions. The Templars' response was unambiguous: "So long as thou dost exercise justice thou wilt reign; but if thou infringe it, thou wilt cease to be King!"

[P4] Ultimately, the impoverished Philip the Fair of France joined forces with Pope Clement V to engineer the Templars' downfall beginning in 1307. Conspiring to seize the order's wealth, Clement invited the Templar Master, Jacques de Molay, to meet with him on the pretext of organizing a new crusade to retake the Holy Land. Shortly thereafter, Philip's forces arrested de Molay and his knights on charges the preponderance of which were almost certainly fabricated. Ranging from the mundane to the unspeakably perverse, the accusations even included an incredible entry citing "every crime and abomination that can be committed." Suffering tortures nearly as horrific as the acts of which they were accused, many Templars confessed.

[P5] Not everyone believed these charges. Despite Philip's urging, Edward II of England remained convinced that the accusations were false, a view seemingly shared by other rulers. Nevertheless, Clement ordered Edward to extract confessions, a task the king tried to carry out with some measure of mercy. By 1313, the Templar Order had been dissolved by papal decree, and many of its members were dead. The following year, Jacques de Molay was burned at the stake after declaring that the Templars' confessions were lies obtained under torture. Stories would spread that as he died, he condemned Clement and Philip to join him within a year. Although it is impossible to say whether he truly called divine vengeance down upon them, within a few months' time both pope and king had gone to their graves.

The Knights Templar ©UWorld

Paragraph 3 Summary

The Templars' growing power led to suspicion and tension with rulers, whose authority they were strong enough to challenge.

Paragraph 4 Summary

The Templars met their end after being tricked by the pope and arrested on false charges by the king of France.

Paragraph 5 Summary

Many rulers believed the accusations were false, but the pope forced them to comply with the Templars' condemnation. Thus, the conflict between the Templars and established authorities ended with the destruction of the Templars by Pope Clement and King Philip of France.

Passage A: The Knights Templar

Content Annotations

Passage Observations: 10 Annotations

[P1] The seal of the Knights Templar depicts two knights astride a single horse, [1] a visual testament of the order's poverty at its inception in 1119. Nevertheless, these Knights of the Temple—who swore oaths not only of poverty but of chastity, loyalty, and bravery—would eventually become one of the wealthiest and most powerful organizations in the medieval world. [2] So far-reaching was their strength and acclaim that their destruction must have seemed as sudden and surprising as it was utter and irrevocable. The signs of danger could not have been wholly invisible, however, as the Templars' growing influence became perceived as a threat to European rulers.

[P2] The first Templars were nine knights who took an oath to defend the Holy Land and any pilgrims who journeyed there after the First Crusade. Having secured a small benefice from Jerusalem's King Baldwin II, the knights inaugurated their mission at the site of the great Temple of Solomon. [3] They quickly attracted widespread admiration as well as many recruits from crusaders and other knights. Within a year of its founding, the order received a financial endowment from the deeply impressed Count of Anjou, whose example was soon followed by other nobles and monarchs. [4] As early as 1128, the Templars even gained official papal recognition, and their wealth, holdings, and numbers swelled both in the Holy Land and throughout Europe, especially in France and England.

[P3] [5] However, this growing power contained the seeds of the order's downfall. Although the Templars were generally held in high esteem, the passage of time saw censure and suspicion directed toward them. The failure of the disastrous siege of Ascalon in 1153 was attributed by some to Templar greed. Similarly, in 1208 Pope Innocent III condemned the wickedness he believed to exist within their ranks. Moreover, their increasingly elevated status brought them into conflict with established authorities. [6] One revealing example occurred in 1252 when, because of the Templars' "many liberties" as well as their "pride and haughtiness," England's King Henry III proposed to curb the order's strength by reclaiming some of its

---

[1] **Claim**

The passage sets up a contrast between the Templars' initial poverty and their eventual wealth and power.

[2] **Focus**

The passage will likely center on explaining the rise and fall of the Knights Templar.

[3] **Claim**

Admiration for the Templars led to voluntary financial support and enthusiastic recruits.

[4] **Elaboration**

Not only did the Templars grow in wealth and power, they did so quickly: their resources and membership swelled less than ten years after their founding.

[5] **Claim**

The author connects the Templars' eventual downfall with their own great power.

[6] **Emphasis**

The author highlights this example as especially representative of how the Templars came into conflict with established rulers. Henry will "cease to be king" if he tries to reclaim the Templars' possessions.

possessions. The Templars' response was unambiguous: "So long as thou dost exercise justice thou wilt reign; but if thou infringe it, thou wilt cease to be King!"

[P4] [7] Ultimately, the impoverished Philip the Fair of France joined forces with Pope Clement V to engineer the Templars' downfall beginning in 1307. Conspiring to seize the order's wealth, Clement invited the Templar Master, Jacques de Molay, to meet with him on the pretext of organizing a new crusade to retake the Holy Land. Shortly thereafter, Philip's forces arrested de Molay and his knights on [8] charges the preponderance of which were almost certainly fabricated. Ranging from the mundane to the unspeakably perverse, the accusations even included an incredible entry citing "every crime and abomination that can be committed." Suffering tortures nearly as horrific as the acts of which they were accused, many Templars confessed.

[P5] [9] Not everyone believed these charges. Despite Philip's urging, Edward II of England remained convinced that the accusations were false, a view seemingly shared by other rulers. Nevertheless, Clement ordered Edward to extract confessions, a task the king tried to carry out with some measure of mercy. By 1313, the Templar Order had been dissolved by papal decree, and many of its members were dead. The following year, Jacques de Molay was burned at the stake after declaring that the Templars' confessions were lies obtained under torture. Stories would spread that as he died, he condemned Clement and Philip to join him within a year. [10] Although it is impossible to say whether he truly called divine vengeance down upon them, within a few months' time both pope and king had gone to their graves.

The Knights Templar ©UWorld

[7] Claim
The Templars' destruction was caused by the pope and the king of France teaming up to ruin them and take their wealth.

[8] Author's View
The author believes that most of the charges against the Templars were false, and that their confessions were forced.

[9] Claim
The Templars' arrest was controversial, as many rulers didn't believe the accusations. Nevertheless, Pope Clement forced these rulers to comply with the arrests and convictions.

[10] Claim
The Templars may have achieved some vengeance for their destruction as both Pope Clement and King Philip soon died.

Passage B: Jackie in 500 Words

This passage is associated with 5 questions analyzed in the UWorld MCAT CARS book.

- Subskill 1a. Main Idea or Purpose Question 2
- Subskill 1b. Meaning of Term Question 3
- Subskill 1c. Direct Passage Claims Question 4
- Subskill 1d. Implicit Claims or Assumptions Question 1
- Subskill 3a. Exemplar Scenario for Passage Claims Question 1

Passage B: Jackie in 500 Words

Structural Annotations

Passage Structure: 5 Annotations

[P1] Born in New York in 1929, Jacqueline Bouvier first came into the public eye as the wife of the 35th president of the United States, John F. Kennedy. The president was assassinated in 1963, and by the end of what turned out to be a turbulent decade, Mrs. Kennedy had transformed herself into the enigmatic Jackie O., wife of Greek shipping magnate Aristotle Onassis. Multifaceted and always elusive, the former first lady never ceased to fascinate; however, people had to be satisfied with only glimpses of this fashion icon, culture advocate, historic preservationist, polyglot, equestrienne, and book editor. Indeed, upon her death in 1994, Jacqueline Kennedy Onassis was described as "the most intriguing woman in the world." Often topping lists of the most admired individuals of the second half of the 20th century, this celebrated woman is likely someone many wish they had known. Barring such a possibility, the best way to fully appreciate Jackie's exceptional nature might be to consider the people she wished she had known.

Paragraph 1 Summary

We can come to understand the "multifaceted" and "fascinating" Jackie better by looking at the people she wished she could have known.

[P2] It is unsurprising that this woman who captured the public's imagination for decades distinguished herself from her peers early on. Notably, in 1951, Ms. Bouvier entered a scholarship contest sponsored by *Vogue* and open to young women in their final undergraduate year, the annual Prix de Paris. Among other assignments, applicants were asked to compose a 500-word essay, "People I Wish I Had Known," spotlighting three individuals influential in art, literature, or culture. The future first lady chose an iconoclastic trio from the Victorian era: the French symbolist poet Charles Baudelaire, the Irish wit Oscar Wilde, and the innovative Ballets Russes dance company founder Sergei Diaghilev.

Paragraph 2 Summary

The author expands on the previous idea of knowing Jackie better by introducing her scholarship essay about people she wished she had known.

[P3] In a brief composition, Jackie provided deep insights into this bohemian threesome of poet, aesthete, and impresario with whom she strongly identified. She concluded that Baudelaire deployed "venom and despair" as "weapons" in his poetry. She idolized Wilde for being able "with the flash of an epigram to bring about what serious reformers had for years been trying to accomplish." Diaghilev she defined as an artist of a different sort, someone who "possessed what is rarer than artistic genius in any one field—the sensitivity to take the best of each man and incorporate it into a masterpiece." As Jackie poignantly observed, such a work is "all the more precious because it lives only in the minds of those who have seen it," dissipating soon after. Furthermore, although these men espoused different disciplines, she discerned that "a common theory runs through their work, a certain concept of the interrelation of the arts." Finally, foreshadowing her self-assumed role in the White House as the nation's unofficial minister of the arts, Jackie paid homage with her vision: "If I could be a sort of Overall Art Director of the Twentieth Century, watching everything from a chair hanging in space, it is their theories of art that I would apply to my period."

Paragraph 3 Summary

Jackie's scholarship essay expressed her admiration for Baudelaire, Wilde, and Diaghilev. All three saw the arts as interconnected, and Jackie would apply their theories to art direction if she had the chance.

[P4] The contest committee judged Jackie's essay to have exhibited a profound appreciation for the arts combined with a truly outstanding level of intellectual maturity and originality of thought. Similarly, biographer Donald Spoto deemed Jackie "remarkably unorthodox," not unlike the men about whom she wrote in her unusual composition which he pronounced "a masterpiece of perceptive improvisation." Thus, from a pool of 1,279 applicants representing 224 colleges, Jacqueline Bouvier was declared the winner.

Paragraph 4 Summary

The author describes why Jackie won the scholarship contest and notes how she and the men she wrote about shared an unorthodox perspective.

[P5] Although Ms. Bouvier went on to decline the prestigious award, which would have involved living and working in Paris, she never gave up her dream of being the century's art director. As first lady, she tirelessly promoted the arts and culture. Today, the John F. Kennedy Center for the Performing Arts in Washington, DC, is a legacy of Jackie's vision.

Paragraph 5 Summary

The vision expressed in Jackie's winning essay informed her adult life and motivated her activities as first lady.

Jackie in 500 Words ©UWorld

Passage B: Jackie in 500 Words

Content Annotations

Passage Observations: 11 Annotations

[P1] Born in New York in 1929, Jacqueline Bouvier first came into the public eye as the wife of the 35th president of the United States, John F. Kennedy. The president was assassinated in 1963, and by the end of what turned out to be a turbulent decade, Mrs. Kennedy had transformed herself into the enigmatic Jackie O., wife of Greek shipping magnate Aristotle Onassis. [1] Multifaceted and always elusive, the former first lady never ceased to fascinate; however, people had to be satisfied with only glimpses of this fashion icon, culture advocate, historic preservationist, polyglot, equestrienne, and book editor. Indeed, upon her death in 1994, Jacqueline Kennedy Onassis was described as [2] "the most intriguing woman in the world." Often topping lists of the most admired individuals of the second half of the 20th century, this celebrated woman is likely someone many wish they had known. Barring such a possibility, [3] the best way to fully appreciate Jackie's exceptional nature might be to consider the people she wished she had known.

[1] Author's View

The author portrays Jackie as being fascinating because of her mysterious nature and many diverse roles in life.

[2] Author's View

By referring to "Jackie's exceptional nature," the author implies that they share the positive attitude many people have toward the "intriguing" and "admired" Jackie.

[3] Focus

The author's goal seems to be to discuss what they see as Jackie's admirable qualities by considering the qualities that Jackie herself admired in others.

[P2] It is unsurprising that this woman who captured the public's imagination for decades distinguished herself from her peers early on. Notably, in 1951, Ms. Bouvier entered a scholarship contest sponsored by *Vogue* and open to young women in their final undergraduate year, the annual Prix de Paris. Among other assignments, applicants were asked to compose a 500-word essay, "People I Wish I Had Known," spotlighting three individuals influential in art, literature, or culture. [4] The future first lady chose an iconoclastic trio from the Victorian era: the French symbolist poet Charles Baudelaire, the Irish wit Oscar Wilde, and the innovative Ballets Russes dance company founder Sergei Diaghilev.

[4] Claim

Jackie's essay centered on three men from the Victorian era: a poet (Baudelaire), a "wit" (Wilde), and the founder of a ballet company (Diaghilev).

[P3] In a brief composition, Jackie provided deep insights into this bohemian threesome of poet, aesthete, and impresario with whom she strongly identified. [5] She concluded that Baudelaire deployed "venom and despair" as "weapons" in his poetry. She idolized Wilde for being able "with the flash of an epigram to bring about what serious reformers had for years been trying to accomplish." Diaghilev she defined as an artist of a different

[5] Outside Source

The passage cites Jackie's own words to describe the specific reasons she admired the men about whom she wrote.

sort, someone who "possessed what is rarer than artistic genius in any one field—the sensitivity to take the best of each man and incorporate it into a masterpiece." As [6] Jackie poignantly observed, such a work is "all the more precious because it lives only in the minds of those who have seen it," dissipating soon after. Furthermore, [7] although these men espoused different disciplines, she discerned that "a common theory runs through their work, a certain concept of the interrelation of the arts." Finally, foreshadowing her self-assumed role in the White House as the nation's unofficial minister of the arts, Jackie paid homage with her vision: [8] "If I could be a sort of Overall Art Director of the Twentieth Century, watching everything from a chair hanging in space, it is their theories of art that I would apply to my period."

[P4] [9] The contest committee judged Jackie's essay to have exhibited a profound appreciation for the arts combined with a truly outstanding level of intellectual maturity and originality of thought. [10] Similarly, biographer Donald Spoto deemed Jackie "remarkably unorthodox," not unlike the men about whom she wrote in her unusual composition which he pronounced "a masterpiece of perceptive improvisation." Thus, from a pool of 1,279 applicants representing 224 colleges, Jacqueline Bouvier was declared the winner.

[P5] Although Ms. Bouvier went on to decline the prestigious award, which would have involved living and working in Paris, [11] she never gave up her dream of being the century's art director. As first lady, she tirelessly promoted the arts and culture. Today, the John F. Kennedy Center for the Performing Arts in Washington, DC, is a legacy of Jackie's vision.

Jackie in 500 Words ©UWorld

---

[6] **Author's View**

The positive term "poignantly" suggests that the author agrees with Jackie's sentiment.

[7] **Likely Question Topic**

The interrelation of the arts is presented as a unifying theme between the different men Jackie admired. Thus, this connection seems important for understanding Jackie's views and could easily serve as the basis for a question.

[8] **Outside Source**

The author again cites Jackie's own words to describe her essay and her overall view of art.

[9] **Claim**

Jackie won the contest because of her maturity, originality, and appreciation of art.

[10] **Outside Source**

The quotations from Donald Spoto largely echo the contest committee's perspective, reinforcing the idea that Jackie's essay displayed unusual insight and quality.

[11] **Claim**

The passage seems to imply that the values expressed in Jackie's essay informed her goals and activities as a promoter of the arts throughout her life.

Passage C: Probability and The Universe

This passage is associated with 6 questions analyzed in the UWorld MCAT CARS book.

- Subskill 1a. Main Idea or Purpose Question 3
- Subskill 1e. Identifying Passage Perspectives Question 1
- Subskill 2a. Logical Relationships Within Passage Question 1
- Subskill 2b. Function of Passage Claim Question 2
- Subskill 2e. Determining Passage Perspectives Question 5
- Subskill 3a. Exemplar Scenario for Passage Claims Question 3

Passage C: Probability and The Universe

Structural Annotations

Passage Structure: 11 Annotations

[P1] The idea of probability is frequently misunderstood, in large part because of a conceptual confusion between objective probability and subjective probability. The failure to make this distinction leads to an erroneous conflation of genuine possibility with what is in fact merely personal ignorance of outcome. An example will clarify.

Paragraph 1 Summary

The passage distinguishes between objective and subjective probability and will go on to illustrate how they are often misunderstood.

[P2] A standard die is rolled on a table, but the outcome of the roll is concealed. Should an observer be asked the chance that a particular number was rolled—five, say—the natural response is 1/6. However, this answer is incorrect. To say there is a one-in-six chance that a five was rolled implies there is an equal chance that any of the other numbers were rolled. But there is no equal chance, because the roll has already occurred. Hence, the probability that the result of the roll is a five is either 100% or 0%, and the same is true for each of the other numbers.

Paragraph 2 Summary

An example of a die roll illustrates the difference between objective and subjective probability: the actual chance of something occurring is unaffected by what a person knows (or doesn't know) about its occurrence.

[P3] [1] It might be objected that such an analysis is an issue of semantics rather than a substantive claim. For declaring the probability to be 1/6 is merely an expression that, for all we know, any number from 1 through 6 might have been rolled. But the difference between *for all we know* and what *is* remains crucially important, because it forestalls a tendency to make scientific assertions from a perspective biased by human perception....

[1] **Addressing Opposing View**

The author anticipates a possible objection to their position so they can refute that objection.

Paragraph 3 Summary

The objective/subjective probability distinction is not just a matter of wording but an important understanding to prevent human bias from influencing scientific assertions.

[P4] [2] [T]hus, it should be clear that claims about the purpose of the universe rest on shaky ground. In particular, we must be wary of inferences drawn from juxtaposing the existence of intelligent life with the genuinely improbable cosmological conditions that make such life possible. Joseph Zycinski provides an instructive account of those conditions: "If twenty billion years ago the rate of expansion were minimally smaller, the universe would have entered a stage of contraction at a time when temperatures of thousands of degrees were prevalent and life could not have appeared. If the distribution of matter were more uniform the galaxies could not have formed. If it were less uniform, instead of the presently

[2] **Topic Shift**

The author switches topics here, building on the "crucially important" distinction in the preceding paragraph to discuss claims about the purpose of the universe.

observed stars, black holes would have formed from the collapsing matter." In short, the existence of life (let alone intelligent life) depends upon the initial conditions of the universe having conformed to an extremely narrow range of possible values.

[P5] [3] It is tempting to go beyond Zycinski's factual point to draw deep cosmological, teleological, and even theological conclusions. But no such conclusions follow. The reason why a life-sustaining universe exists is that if it did not, there would be no one to wonder why a life-sustaining universe exists. This fact is a function not of purpose, but of pre-requisite. [4] For instance, suppose again that a five was rolled on a die. We now observe a five on its face, not because the five was "meant to be" but because one side of the die had to land face up. Even if we imagine that the die has not six but six *billion* sides, that analysis is unaffected. Yes, it was unlikely that any given value would be rolled, but *some* value had to be rolled. That "initial condition" then set the parameters for what kind of events could possibly follow; in this case, the observation of the five.

[P6] Similarly, the existence of intelligent beings is evidence that certain physical laws obtained in the universe. However, it is not evidence that those beings or laws were necessary or intended rather than essentially random. [5] The subjective probability of that outcome is irrelevant, because in this universe the objective probability of its occurrence is 100%. Thus, the seemingly low initial chance that such a universe would exist is not in itself indicative of a purpose to its existence.

Probability and The Universe ©UWorld

Paragraph 4 Summary

The author applies the previous points about probability and human bias to the existence of the universe. Even though the conditions for life were extremely improbable, that fact does not justify inferences about the universe's purpose.

[3] Addressing Opposing View

The author raises a "tempting" view in order to argue against it.

[4] Connected Ideas

The connection between the original die-rolling example and this new, more extensive example parallels the topic shift from probability to the existence of the universe. The original example illustrated the right way to think about probability; this second example illustrates the right way to think about whether the universe has a purpose.

Paragraph 5 Summary

The conditions necessary for intelligent life were improbable, but that doesn't mean they were "meant to be." A die with 6 billion sides is an analogy—any individual roll is unlikely, but some side has to land face up.

[5] Connected Ideas

The author directly links their claims about the purpose of the universe to the distinction in Paragraph 1 between objective and subjective probability.

Paragraph 6 Summary

The distinction between objective and subjective probability parallels the difference between facts and ideas about purpose. Just because a universe with intelligent life was improbable does not mean there is a purpose to its existence.

Passage C: Probability and The Universe

Content Annotations

Passage Observations: 9 Annotations
Potential Pitfalls: 2 Annotations

[P1] [1] The idea of probability is frequently misunderstood, in large part because of a conceptual confusion between objective probability and subjective probability. The failure to make this distinction leads to an erroneous conflation of genuine possibility with what is in fact merely personal ignorance of outcome. An example will clarify.

[1] **Claim**

The passage distinguishes between "objective" and "subjective" probability, which parallels the contrast between "genuine possibility" and "merely personal ignorance." Thus, the implication seems to be that only objective probability is genuine.

[P2] A standard die is rolled on a table, but the outcome of the roll is concealed. Should an observer be asked the chance that a particular number was rolled—five, say—[1] the natural response is 1/6. However, this answer is incorrect. [2] To say there is a one-in-six chance that a five was rolled implies there is an equal chance that any of the other numbers were rolled. But there is no equal chance, because the roll has already occurred. Hence, the probability that the result of the roll is a five is either 100% or 0%, and the same is true for each of the other numbers.

[1] **Potential Pitfall: Trap Answer**

Since the author says this "natural response" is wrong, questions and answer choices might play on the tendency to think in this way in order to trick you.

[2] **Elaboration**

According to the author, this is the correct (but not obvious) way of understanding the probability of the roll. We now have the information that was meant to clarify the author's points from the previous paragraph.

[P3] It might be objected that such an analysis is an issue of semantics rather than a substantive claim. For declaring the probability to be 1/6 is merely an expression that, for all we know, any number from 1 through 6 might have been rolled. But [3] the difference between *for all we know* and what *is* remains crucially important, because it forestalls a tendency to make scientific assertions from a perspective biased by human perception….

[3] **Emphasis**

The author highlights this difference as crucially important, suggesting this point is a key part of their argument.

[P4] [T]hus, it should be clear that claims about the purpose of the universe rest on shaky ground. In particular, we must be wary of inferences drawn from juxtaposing the existence of intelligent life with the genuinely improbable cosmological conditions that make such life possible. [4] Joseph Zycinski provides an instructive account of those conditions: "If twenty billion years ago the rate of expansion were minimally smaller, the universe would have entered a stage of contraction at a time when temperatures of thousands of degrees were prevalent and life could not have appeared. If the distribution of matter were more uniform the galaxies could not have

[4] **Outside Source**

The author cites Zycinski's words to explain the specific conditions necessary for a universe to support intelligent life.

formed. If it were less uniform, instead of the presently observed stars, black holes would have formed from the collapsing matter." [5] In short, the existence of life (let alone intelligent life) depends upon the initial conditions of the universe having conformed to an extremely narrow range of possible values.

[P5] It is tempting to go beyond [6] Zycinski's factual point to draw deep cosmological, teleological, and even theological conclusions. [2] But no such conclusions follow. [7] The reason why a life-sustaining universe exists is that if it did not, there would be no one to wonder why a life-sustaining universe exists. This fact is a function not of purpose, but of prerequisite. For instance, suppose again that a five was rolled on a die. We now observe a five on its face, not because the five was "meant to be" but because one side of the die had to land face up. Even if we imagine that the die has not six but six *billion* sides, that analysis is unaffected. Yes, it was unlikely that any given value would be rolled, but *some* value had to be rolled. That "initial condition" then set the parameters for what kind of events could possibly follow; in this case, the observation of the five.

[P6] [8] Similarly, the existence of intelligent beings is evidence that certain physical laws obtained in the universe. However, it is not evidence that those beings or laws were necessary or intended rather than essentially random. The subjective probability of that outcome is irrelevant, because in this universe the objective probability of its occurrence is 100%. [9] Thus, the seemingly low initial chance that such a universe would exist is not in itself indicative of a purpose to its existence.

Probability and The Universe ©UWorld

---

[5] Elaboration
The passage first provides and then summarizes Zycinski's quote, which explains that the existence of life depends on "extremely narrow" conditions.

[6] Author's View
The author agrees with Zycinski's view, describing it as "factual."

[2] Potential Pitfall: Misreading
It would be easy to think that "no such conclusions follow" means the author is disagreeing with Zycinski. But attentive reading shows the opposite is true—the author calls Zycinski's point "factual" and is cautioning against going *beyond* what he said.

[7] Claim
As a counterpoint to the "tempting" conclusions just mentioned, the author asserts that nothing about purpose can be inferred from the mere fact that intelligent life exists. We can ask why intelligent life exists only because the necessary conditions for it did occur, however improbable those conditions might seem.

[8] Author's View
The author further stresses that the low probability of intelligent life existing is not evidence that such life was meant to exist rather than a random occurrence.

[9] Focus
This concluding sentence likely represents the main point of the passage. The author's argument is meant to establish that: the low subjective probability that a universe with intelligent life would exist is not evidence that there is a purpose to the universe's existence.

Passage D: American Local Motives

This passage is associated with 7 questions analyzed in the UWorld MCAT CARS book.

- Subskill 1b. Meaning of Term Question 1
- Subskill 2a. Logical Relationships Within Passage Question 2
- Subskill 2c. Extent of Passage Evidence Question 1
- Subskill 2d. Connecting Claims With Evidence Question 3
- Subskill 3c. New Claim Support or Challenge Question 4
- Subskill 3d. External Scenario Support or Challenge Question 1
- Subskill 3g. Additional Conclusions From New Information Question 3

Passage D: American Local Motives

Structural Annotations

Passage Structure: 9 Annotations

[P1] Locomotives were invented in England, with the first major railroad connecting Liverpool and Manchester in 1830. However, it was in America that railroads would be put to the greatest use in the nineteenth century. On May 10, 1869, the Union Pacific and Central Pacific lines met at Promontory Point, Utah, joining from opposite directions to complete a years-long project—the Transcontinental Railroad. This momentous event connected the eastern half of the United States with its western frontier and facilitated the construction of additional lines in between. As a result, journeys that had previously taken several months by horse and carriage now required less than a week's travel. By 1887 there were nearly 164,000 miles of railroad tracks in America, and by 1916 that number had swelled to over 254,000.

Paragraph 1 Summary

Trains were invented in nineteenth-century England but used more extensively in America. The completion of the Transcontinental Railroad in 1869 dramatically transformed travel in the United States, and the amount of railroad tracks continued to increase over the next few decades.

[P2] While the United States still has the largest railroad network in the world, it operates largely in the background of American life, and citizens no longer view trains with the sense of importance those machines once commanded. Nevertheless, the economic and industrial advantages those citizens enjoy today would not have been possible without America's history of trains; as Tom Zoellner reminds us, "Under the skin of modernity lies a skeleton of railroad tracks." [1] Although airplanes and automobiles have now assumed greater prominence, the time has arrived for the resurgence of railroads. A revitalized and advanced railway system would confer numerous essential benefits on both the United States and the globe.

[1] **Topic Shift**

While so far the passage has provided historical information about American railroads, the author now switches to arguing that railroads should be increasingly used in the future.

Paragraph 2 Summary

Trains were essential to creating modern American society but have fallen out of prominence in favor of cars and planes. In the author's view, however, they should once again be used to transform American life.

[P3] [2] The chief obstacles to garnering support for such a project are the current dominance of the automobile and the languishing technology of existing railroads. In a sense these two obstacles are one, as American dependence on personal automobiles is partially due to the paucity of rapid public transportation. The railroads of Europe and Japan, by comparison, have vastly outpaced their American counterparts. Japan has operated high-speed rail lines continuously since 1964, and in 2007, a French train set a record of 357 miles per

[2] **Addressing Opposing View**

The author discusses two factors that could inhibit people from supporting a revitalized train system.

hour. While that speed was achieved under tightly controlled conditions, it still speaks to the great disparity in railroad development between the United States and other countries since the mid-twentieth century. British trains travel at speeds much higher than those in America, where both the trains themselves and the infrastructure to support them have simply been allowed to fall behind. In much of Europe it is common for trains to travel at close to 200 miles per hour.

[P4] [3] To invest in a modern network of railroads would improve the United States in much the same way that the first railroads did in the nineteenth and early twentieth centuries. A high-speed passenger rail system would dramatically transform American life, as travel between cities and states would become quicker and more convenient, encouraging commerce, business, and tourism. Such a system would also make important strides in environmental preservation. According to a 2007 British study, "CO2 emissions from aircraft operations are...at least five times greater" than those from high-speed trains. For similar reasons, Osaka, Japan, was ranked as "the best…green transportation city in Asia" by the *2011 Green City Index*. As Lee-in Chen Chiu notes in *The Kyoto Economic Review*, Osakans travel by railway more than twice as much as they travel by car.

[P5] [4] It is true that developing a countrywide high-speed rail system would come with significant costs. However, that was also true of the original Transcontinental Railroad, as indeed it is with virtually any great project undertaken for the public good. We should thus move ahead with confidence that the rewards will outweigh the expenditure as citizens increasingly choose to travel by train. Both for society's gain and the crucial well-being of the planet, our path forward should proceed upon rails.

American Local Motives ©UWorld

Paragraph 3 Summary

Americans today generally prefer automobiles to trains, but this can be attributed to the poor state of American railways compared to other parts of the world. Japan, England, and much of Europe have developed vastly superior modern trains.

[3] Connected Ideas

The author draws an explicit comparison between the proposed modern rail system and the dramatic impact that railroads have historically had in the U.S.

Paragraph 4 Summary

The author argues that high-speed trains would create improvements in many areas of American life. Such trains would also be good for the environment, as indicated by data from multiple studies.

[4] Connected Ideas

As in the previous paragraph, the author compares the historical impact of the first railroads with the potential impact of a new, high-speed rail system. The dramatic transformation such a modern system would create seems to be an overarching theme of the passage.

Paragraph 5 Summary

The author argues that although a high-speed rail system would be expensive, the benefits to society and the environment would outweigh those costs. Just as the Transcontinental Railroad was enormously beneficial in the nineteenth century, we should embrace a modern rail system today.

Passage D: American Local Motives

Content Annotations

Passage Observations: 9 Annotations

[P1] [1] Locomotives were invented in England, with the first major railroad connecting Liverpool and Manchester in 1830. However, it was in America that railroads would be put to the greatest use in the nineteenth century. On May 10, 1869, the Union Pacific and Central Pacific lines met at Promontory Point, Utah, joining from opposite directions to complete a years-long project—the Transcontinental Railroad. [2] This momentous event connected the eastern half of the United States with its western frontier and facilitated the construction of additional lines in between. As a result, journeys that had previously taken several months by horse and carriage now required less than a week's travel. By 1887 there were nearly 164,000 miles of railroad tracks in America, and by 1916 that number had swelled to over 254,000.

[P2] While the United States still has the largest railroad network in the world, it operates largely in the background of American life, and citizens no longer view trains with the sense of importance those machines once commanded. [3] Nevertheless, the economic and industrial advantages those citizens enjoy today would not have been possible without America's history of trains; as Tom Zoellner reminds us, "Under the skin of modernity lies a skeleton of railroad tracks." Although airplanes and automobiles have now assumed greater prominence, the time has arrived for the resurgence of railroads. [4] A revitalized and advanced railway system would confer numerous essential benefits on both the United States and the globe.

[P3] The chief obstacles to garnering support for such a project are the current dominance of the automobile and the languishing technology of existing railroads. [5] In a sense these two obstacles are one, as American dependence on personal automobiles is partially due to the paucity of rapid public transportation. The railroads of Europe and Japan, by comparison, have vastly outpaced their American counterparts. [6] Japan has operated high-speed rail lines continuously since 1964, and in 2007, a French train set a record of 357 miles per

[1] **Claim**

While trains were an English invention, they had more effect on life in America.

[2] **Claim**

The completion of the Transcontinental Railroad is described as "momentous," with a comparison illustrating how drastically it changed the speed of travel.

[3] **Outside Source**

The author quotes Zoellner to support the assertion that trains were essential to creating modern American society.

[4] **Focus**

Based on this sentence and the previous one, it seems the author's main purpose is to argue for the creation of a revitalized American railway system. In the author's view, the restored prominence of trains would produce both domestic and global benefits.

[5] **Author's View**

According to the author, the outdated technology of current railroads and the American preference for automobiles are two sides of the same coin. Thus, the implication seems to be that this preference would change if better railroads were available.

[6] **Elaboration**

Data is given to illustrate the superiority of trains in Europe and Japan compared to the "languishing technology" of trains in America.

hour. While that speed was achieved under tightly controlled conditions, it still speaks to the great disparity in railroad development between the United States and other countries since the mid-twentieth century. British trains travel at speeds much higher than those in America, where both the trains themselves and the infrastructure to support them have simply been allowed to fall behind. In much of Europe it is common for trains to travel at close to 200 miles per hour.

[P4] To invest in a modern network of railroads would improve the United States in much the same way that the first railroads did in the nineteenth and early twentieth centuries. [7] A high-speed passenger rail system would dramatically transform American life as travel between cities and states would become quicker and more convenient, encouraging commerce, business, and tourism. Such a system would also make important strides in environmental preservation. [8] According to a 2007 British study, "CO2 emissions from aircraft operations are...at least five times greater" than those from high-speed trains. For similar reasons, Osaka, Japan, was ranked as "the best...green transportation city in Asia" by the 2011 Green City Index. As Lee-in Chen Chiu notes in *The Kyoto Economic Review*, Osakans travel by railway more than twice as much as they travel by car.

[7] **Author's View**

Although the author has previously mentioned that high-speed trains would be beneficial, they only now state specifically what those benefits would be: improvements in commerce, business, tourism, and the environment.

[8] **Outside Source**

Data from British and Japanese sources support the author's claim that high-speed rail is good for the environment.

[P5] It is true that developing a countrywide high-speed rail system would come with significant costs. However, that was also true of the original Transcontinental Railroad, as indeed it is with virtually any great project undertaken for the public good. We should thus move ahead with [9] confidence that the rewards will outweigh the expenditure as citizens increasingly choose to travel by train. Both for society's gain and the crucial well-being of the planet, our path forward should proceed upon rails.

[9] **Author's View**

According to the author, we can count on people wanting to travel by train once a modern railway is available.

American Local Motives ©UWorld

Passage E: The Divine Sign of Socrates

This passage is associated with 6 questions analyzed in the UWorld MCAT CARS book.

- Subskill 1b. Meaning of Term Question 2
- Subskill 1d. Implicit Claims or Assumptions Question 2
- Subskill 2e. Determining Passage Perspectives Question 3
- Subskill 2e. Determining Passage Perspectives Question 4
- Subskill 2f. Drawing Additional Inferences Question 2
- Subskill 3b. Passage Applications to New Context Question 2

Passage E: The Divine Sign of Socrates

Structural Annotations

Passage Structure: 5 Annotations

[P1] From his bare feet to his bald pate, the potentially shapeshifting figure of Socrates found in the literary tradition that arose after his controversial trial and death presents an intriguing array of oddities and unorthodoxies. Most conspicuously, his unshod and shabby sartorial state flaunted poverty at a time when the city of Athens had become obsessed with wealth and its trappings. Yet the philosopher's peculiar appearance was but a hint of the strange new calling he embraced. Inspired perhaps by the famous Delphic dictum "Know thyself," he embarked on a mission devoted to finding truth through dialogue. In what struck some as a dangerous new method of inquiry, he subjected nearly everyone he encountered to intense cross-examination, mercilessly exposing the ignorance of his interlocutors. Moreover, in a culture that still put stock in magic, the highly charismatic, entertaining, and at times infuriating Socrates appeared to be a sorcerer bewitching the aristocratic young men of Athens who followed him fanatically about the agora.

Paragraph 1 Summary

Socrates was a strange figure whose odd behavior, constant questioning, and bewitching influence over the youth caused some to view him with alarm.

[P2] By all credible accounts, this exceedingly eccentric, self-styled radical truth-seeker had more than a whiff of the uncanny about him. As Socrates himself explains in Plato's *Republic*, he was both blessed and burdened with a supernatural phenomenon in the form of a *daimonion* or inner spirit that always guided him: "This began when I was a child. It is a voice, and whenever it speaks, it turns me away from something I am about to do, but it never encourages me to do anything." An overtly rational thinker, Socrates nonetheless considered these warnings—or, in James Miller's words, "the audible interdictions he experienced as irresistible"—to be infallible. Such oracular injunctions were highly anomalous as tutelary spirits were thought to assume a more nuanced presence. Some scholars have dismissed Socrates' recurring sign as a hallucination or psychological aberration. Others have conjectured that the internal voice might be attributable to the cataleptic or trancelike episodes from which the philosopher purportedly suffered. Indeed, as Miller notes, "Socrates was

Paragraph 2 Summary

A particularly odd feature of Socrates was the supposed inner voice that he claimed guided him. Such a voice did not fit with typical storied for the abstracted states that overtook him"; not infrequently, his companions would see him stop in his tracks and stand still for hours, completely lost in thought.

Athenian belief, and some modern scholars attribute it to psychological causes. However, Socrates viewed it as an infallible spirit that gave him warnings.

[P3] As Socrates further insisted, it was only the protestations of this apotreptic voice that held him back from entering the political arena. Even so, its personal admonitions could not spare him persecution. Despite the political amnesty extended by the resurgent democracy that succeeded the interim pro-Spartan oligarchy, the thinker's notoriety and ambiguous allegiances aroused suspicions. In 399 BCE, Socrates was brought before the court on trumped-up charges of impiety; these included [1] willfully neglecting the traditional divinities, flagrantly introducing new gods to the city, and wittingly corrupting the youth. Athenian society recognized no division between religious and civic duties, and capricious gods demanded constant appeasement through sacrifices and rituals. Consequently, belief in a purely private deity—particularly a wholly benevolent deity conveying unequivocal messages—was inadmissible. Worse, as Socrates' own testimony revealed, he honored this personal god's authority above even the laws of the city. Hence, the philosopher's *daimonion* loomed over his indictment, conviction, and sentencing.

[1] **Connected Ideas**

Here we can see a thread connecting the first three paragraphs. Socrates seemed to "bewitch" the young men of Athens, and he spoke of a strange spirit that guided him. Thus, he was accused of corrupting the youth and introducing new gods to the city.

Paragraph 3 Summary

Socrates' *daimonion* was used as an excuse to bring him to trial on politically motivated charges. Nevertheless, Socrates admitted he saw his inner spirit as a higher authority than the city's laws. Thus, the *daimonion* contributed to his conviction for impiety and corrupting the youth.

[P4] Nevertheless, in his defense speech as reconstructed by Plato in the *Apology*, Socrates maintained confidence in the protective nature and prophetic powers of his inner monitor. He never questioned its affirmatory silence toward his predicament, remarking, "The divine faculty would surely have opposed me had I been going to evil and not to good." Thus, Socrates acknowledged that his *daimonion* had its reasons, however inscrutable. Variously described as malcontent and martyr, public nuisance and prophet, laughingstock and hero, the mercurial Athenian, like the sign that guided him, was difficult to fathom yet impossible to ignore.

Paragraph 4 Summary

Socrates continued to believe his *daimonion* was infallible even after he was convicted and sentenced. Thus, this guiding spirit represented the remarkable strangeness of Socrates' life and character.

The Divine Sign of Socrates ©UWorld

Passage E: The Divine Sign of Socrates

Content Annotations

Passage Observations: 12 Annotations
Potential Pitfalls: 1 Annotation

[P1] From his bare feet to his bald pate, [1] the potentially shapeshifting figure of Socrates found in the literary tradition that arose after his controversial trial and death presents an intriguing array of oddities and unorthodoxies. Most conspicuously, his unshod and shabby sartorial state flaunted poverty at a time when the city of Athens had become obsessed with wealth and its trappings. [2] Yet the philosopher's peculiar appearance was but a hint of the strange new calling he embraced. Inspired perhaps by the famous Delphic dictum "Know thyself," he embarked on a mission devoted to finding truth through dialogue. In what struck some as a dangerous new method of inquiry, he subjected nearly everyone he encountered to intense cross-examination, mercilessly exposing the ignorance of his interlocutors. Moreover, [3] in a culture that still put stock in magic, the highly charismatic, entertaining, and at times infuriating Socrates appeared to be a sorcerer bewitching the aristocratic young men of Athens who followed him fanatically about the agora.

[P2] By all credible accounts, [4] this exceedingly eccentric, self-styled radical truth-seeker had more than a whiff of the uncanny about him. As Socrates himself explains in Plato's *Republic*, he was both blessed and burdened with a supernatural phenomenon in the form of a *daimonion* or inner spirit that always guided him: [5] "This began when I was a child. It is a voice, and whenever it speaks, it turns me away from something I am about to do, but it never encourages me to do anything." An overtly rational thinker, Socrates nonetheless considered these warnings—or, in James Miller's words, "the audible interdictions he experienced as irresistible"—to be infallible. Such oracular injunctions were [6] highly anomalous as tutelary spirits were thought to assume a more nuanced presence. Some scholars have dismissed Socrates' recurring sign as a hallucination or psychological aberration. Others have conjectured that the internal voice might be attributable to [1] the cataleptic or trancelike episodes from which the

---

[1] **Author's View**

The author portrays Socrates as mysterious and strange.

[2] **Claim**

Socrates' behavior was even more disconcerting than his appearance, as he habitually questioned everyone he met in an attempt to discover truth.

[3] **Claim**

Socrates drew people to himself easily, but some found his behavior and sway over the youth to be a cause for anger or concern.

[4] **Focus**

The author moves from an overview of Socrates' strangeness to discussing a specific oddity in depth. Thus, Socrates' *daimonion* is likely the main subject of the passage. (This inference is also suggested by the passage's title.)

[5] **Outside Source**

Both Socrates himself and James Miller are quoted to support the passage description of Socrates' *daimonion*.

[6] **Claim**

Although Socrates saw his *daimonion*'s warnings as both clear and infallible, this view diverged from typical Athenian belief about spirits.

[1] **Potential Pitfall: Misreading**

The author affirms that Socrates experienced trancelike episodes but does not otherwise

philosopher purportedly suffered. Indeed, as Miller notes, "Socrates was storied for the abstracted states that overtook him"; not infrequently, his companions would see him stop in his tracks and stand still for hours, completely lost in thought.

[P3] [7] As Socrates further insisted, it was only the protestations of this apotreptic voice that held him back from entering the political arena. Even so, its personal admonitions could not spare him persecution. Despite the political amnesty extended by the resurgent democracy that succeeded the interim pro-Spartan oligarchy, [8] the thinker's notoriety and ambiguous allegiances aroused suspicions. In 399 BCE, Socrates was brought before the court on trumped-up charges of impiety; these included willfully neglecting the traditional divinities, flagrantly introducing new gods to the city, and wittingly corrupting the youth. [9] Athenian society recognized no division between religious and civic duties, and capricious gods demanded constant appeasement through sacrifices and rituals. Consequently, belief in a purely private deity—particularly a wholly benevolent deity conveying unequivocal messages—was inadmissible. [10] Worse, as Socrates' own testimony revealed, he honored this personal god's authority above even the laws of the city. Hence, the philosopher's *daimonion* loomed over his indictment, conviction, and sentencing.

[P4] Nevertheless, in his defense speech as reconstructed by Plato in the *Apology*, [11] Socrates maintained confidence in the protective nature and prophetic powers of his inner monitor. He never questioned its affirmatory silence toward his predicament, remarking, "The divine faculty would surely have opposed me had I been going to evil and not to good." [12] Thus, Socrates acknowledged that his *daimonion* had its reasons, however inscrutable. Variously described as malcontent and martyr, public nuisance and prophet, laughingstock and hero, the mercurial Athenian, like the sign that guided him, was difficult to fathom yet impossible to ignore.

The Divine Sign of Socrates ©UWorld

---

express agreement or disagreement with the scholars mentioned. Thus, we cannot necessarily infer the author's own view about Socrates' *daimonion*.

[7] Claim
The voice warned Socrates not to get involved in politics. Despite following this advice, Socrates was eventually persecuted anyway.

[8] Author's View
According to the author, the charges against Socrates were unjust and politically motivated.

[9] Likely Question Topic
Here we have specific information about why Socrates and his *daimonion* seemed troubling to Athenian society. It would not be surprising to see a question asking about the reasons for Socrates' trial or why his actions were viewed as disruptive.

[10] Emphasis
There were many reasons why the Athenians viewed Socrates' *daimonion* with suspicion, but especially concerning was that Socrates respected it more than he did the city laws.

[11] Claim
Socrates had faith in his *daimonion* even after his conviction, believing that things must have turned out for the best.

[12] Focus
The main purpose of the passage seems to be to describe Socrates' *daimonion* and its role in his unusual life.

Passage F: The Victorian Internet

This passage is associated with 5 questions analyzed in the UWorld MCAT CARS book.

- Subskill 1c. Direct Passage Claims Question 1
- Subskill 1d. Implicit Claims or Assumptions Question 4
- Subskill 1e. Identifying Passage Perspectives Question 2
- Subskill 1f. Further Implications of Passage Claims Question 2
- Subskill 2b. Function of Passage Claim Question 1

Passage F: The Victorian Internet

Structural Annotations

Passage Structure: 8 Annotations

[P1] Lasting roughly from 1820 to 1914, the Victorian Era is often defined by its many distinctive sociological conditions, including industrialization, urbanization, railroad travel, imperialism, territorial expansionism, and the frictions sparked by Darwinism and democratic reform. However, descriptors that may not come as readily to mind are "the Information Age" and "the Age of Communication"; nor would we likely associate this era with something called the "highway of thought," which emanated from the electric telegraph—an invention that, in retrospect, could be renamed "the Victorian Internet." Yet the rise of the electric telegraph in the mid-1800s constituted the greatest revolution in communication since the invention of the printing press in the fifteenth century and until the launch of the World Wide Web at the end of the twentieth century.

Paragraph 1 Summary

The Victorian Era is not usually thought of as an "Age of Communication" but it should be; the invention of the electric telegraph was a revolutionary event in human history.

[P2] Historically, messages could be conveyed only as fast as a person could travel from one location to another. However, by the end of the eighteenth century, the Chappe brothers had constructed a rudimentary optical telegraph on a hilltop tower in France. Using the large, jointed arms attached to the roof, operators could form various configurations to communicate a message to a similar tower farther away, which would then relay the message to a third tower, and so on, in a long-distance chain. Nevertheless, even with telescopes, visibility severely hampered the efficacy of this semaphore system. After countless attempts by scientists to use electricity to transmit messages, innovators in both Britain and America, including the painter and polymath Samuel F.B. Morse, worked at harnessing electromagnetic forces to send communications via cable. But while in Britain the technology was initially reserved for railway signaling, in the U.S. it culminated in the transmission of a message along a 40-mile telegraphic wire from Washington, D.C., to Baltimore in 1844. Using a binary code of short and long electrical impulses, or "dots and dashes," Morse dispatched the words: "What hath God wrought?"

Paragraph 2 Summary

An earlier optical messaging system had been used in France, but it was limited and impractical. By the mid-eighteenth century, British and American inventors like Morse succeeded in sending messages by electrical wire.

[P3] Within 20 years, telegraph cables crisscrossed the continental U.S. and much of Europe. The telegraph office became ubiquitous, and [1] telegrams—often bouncing from one office to another like emails tossed from server to server—reached ever more remote destinations. Following some spectacular failures, a durable cable was stretched across the floor of the Atlantic Ocean, enabling Europe and the U.S. to exchange messages within minutes. In the 1870s, the British Empire connected London with outposts in India and Australia. While individual telegram speed remained relatively constant, an endless deluge of information began to pour through the wires. There was, as Tom Standage observes, "an irreversible acceleration in the pace of business life," reflecting how "telegraphy and commerce thrived in a virtuous circle." It was as though "rapid long-distance communication had effectively obliterated time and space," begetting the phenomenon known as "globalization."

[P4] A new type of skilled worker, the telegrapher, was born. He or she belonged to a vast, online community, whose semi-anonymous members shared a unique intermediary role as well as a language of dots and dashes—[2] vaguely prefiguring the exchange of bits and bytes along modern computer networks. Like today's online communities, these telegraphers fostered their own subculture: jokes and anecdotes flew over the wires; some operators invented systems to play games, and occasional romances blossomed. Regrettably, hackers, scammers, and shady entrepreneurs also frequented the byways of this early internet.

[P5] Predictably, newer forms of the original telegraphic technology—first the telephone and, much later, the fax machine—eventually encroached. The last telegram departed from India in 2013, but [3] the twenty-first century has arguably seen a revival of this communication form in the text message. As Standage claims, texting has not only "reincarnate[d] a defunct nineteenth-century technology" but reinforced "the democratization of telecommunications" inaugurated by the miraculous Victorian Internet.

The Victorian Internet ©UWorld

---

[1] Connected Ideas

This comparison extends the analogy presented in Paragraph 1 between the telegraph and the modern internet.

Paragraph 3 Summary

Telegraphic communication increased rapidly both domestically and intercontinentally. Communication and business mutually thrived, giving rise to globalization.

[2] Connected Ideas

The passage draws a further analogy between the "Victorian Internet" and the modern internet.

Paragraph 4 Summary

Telegraphers were an important new type of worker, and they took part in online communities that were similar in many ways to today's internet.

[3] Connected Ideas

The passage again describes how the telegraph was like the modern internet, here directly comparing telegrams with text messages. This extended comparison is clearly an overarching theme of the passage.

Paragraph 5 Summary

Although there are no more telegrams now, text messages are arguably a modern incarnation of them. Thus, the communication revolution begun by the telegraph continues today.

Passage F: The Victorian Internet

Content Annotations

Passage Observations: 7 Annotations
Potential Pitfalls: 1 Annotation

[P1] Lasting roughly from 1820 to 1914, the Victorian Era is often defined by its many distinctive sociological conditions, including industrialization, urbanization, railroad travel, imperialism, territorial expansionism, and the frictions sparked by Darwinism and democratic reform. [1] However, descriptors that may not come as readily to mind are "the Information Age" and "the Age of Communication"; nor would we likely associate this era with something called the "highway of thought," which emanated from the electric telegraph—[2] an invention that, in retrospect, could be renamed "the Victorian Internet." Yet the rise of the electric telegraph in the mid-1800s constituted the greatest revolution in communication since the invention of the printing press in the fifteenth century and until the launch of the World Wide Web at the end of the twentieth century.

[1] **Author's View**

While the Victorian Era is known for many things, the author believes the significance of the electric telegraph is underacknowledged.

[2] **Focus**

Given the term "revolution in communication," the passage will likely center on explaining the telegraph's immense societal impact.

[P2] Historically, messages could be conveyed only as fast as a person could travel from one location to another. However, by the end of the eighteenth century, the Chappe brothers had constructed a rudimentary optical telegraph on a hilltop tower in France. Using the large, jointed arms attached to the roof, operators could form various configurations to communicate a message to a similar tower farther away, which would then relay the message to a third tower, and so on, in a long-distance chain. [3] Nevertheless, even with telescopes, visibility severely hampered the efficacy of this semaphore system. [1] After countless attempts by scientists to use electricity to transmit messages, innovators in both Britain and America, including the painter and polymath Samuel F.B. Morse, worked at harnessing electromagnetic forces to send communications via cable. But while in Britain the technology was initially reserved for railway signaling, in the U.S. it culminated in the transmission of a message along a 40-mile telegraphic wire from Washington, D.C., to Baltimore in 1844. Using a binary code of short and long electrical impulses, or "dots and dashes," Morse dispatched the words: "What hath God wrought?"

[3] **Claim**

Previous attempts at long-distance communication were still limited by dependence on proximity and vision.

[1] **Potential Pitfall: Details**

These sentences contain nuances that could lead to wrong answers. For instance, the passage says the technology was "initially reserved" for railway signaling in Britain, but it doesn't say for how long, nor whether the U.S. also used it for railways in addition to messages.

[P3] Within 20 years, telegraph cables crisscrossed the continental U.S. and much of Europe. The telegraph office became ubiquitous, and telegrams—often bouncing from one office to another like emails tossed from server to server—reached ever more remote destinations. Following some spectacular failures, a durable cable was stretched across the floor of the Atlantic Ocean, enabling Europe and the U.S. to exchange messages within minutes. In the 1870s, the British Empire connected London with outposts in India and Australia. While individual telegram speed remained relatively constant, [4] an endless deluge of information began to pour through the wires. There was, as Tom Standage observes, "an irreversible acceleration in the pace of business life," reflecting how "telegraphy and commerce thrived in a virtuous circle." It was as though [5] "rapid long-distance communication had effectively obliterated time and space," begetting the phenomenon known as "globalization."

[4] **Outside Source**

Standage's quote describes a mutually reinforcing relationship between telegraphy and commerce in which each contributed to the rapid growth of the other.

[5] **Likely Question Topic**

Given how dramatically the electric telegraph changed the world, a question could easily ask about its effects or how life differed before and after its introduction.

[P4] A new type of skilled worker, the telegrapher, was born. He or she belonged to a vast, online community, whose semi-anonymous members shared a unique intermediary role as well as a language of dots and dashes—vaguely prefiguring the exchange of bits and bytes along modern computer networks. Like today's online communities, these telegraphers fostered their own subculture: jokes and anecdotes flew over the wires; some operators invented systems to play games, and occasional romances blossomed. Regrettably, hackers, scammers, and shady entrepreneurs also frequented the byways of this early internet.

[P5] Predictably, newer forms of the original telegraphic technology—first the telephone and, much later, the fax machine—eventually encroached. The last telegram departed from India in 2013, but the twenty-first century has arguably seen a revival of this communication form in the text message. [6] As Standage claims, texting has not only "reincarnate[d] a defunct nineteenth-century technology" but reinforced "the democratization of telecommunications" inaugurated by [7] the miraculous Victorian Internet.

[6] **Outside Source**

The author again cites Standage's words to support claims about the impact and importance of the telegraph.

[7] **Author's View**

The author calls the telegraph "miraculous." Along with the previous description of it as the "greatest revolution in communication" between the printing press and the modern internet, the author appears to view the telegraph as both extremely influential and positive.

The Victorian Internet ©UWorld

Passage G: Food Costs and Disease

This passage is associated with 6 questions analyzed in the UWorld MCAT CARS book.

- Subskill 1c. Direct Passage Claims Question 2
- Subskill 1e. Identifying Passage Perspectives Question 4
- Subskill 2b. Function of Passage Claim Question 3
- Subskill 3b. Passage Application to New Context Question 1
- Subskill 3d. External Scenario Support or Challenge Question 2
- Subskill 3e. Applying Passage Perspectives Question 3

Passage G: Food Costs and Disease

Structural Annotations

Passage Structure: 10 Annotations

[P1] Because frequent consumption of unhealthy foods is strongly linked with cardiometabolic diseases, one way for governments to combat those afflictions may be to modify the eating habits of the general public. Applying economic incentives or disincentives to various types of foods could potentially alter people's diets, leading to more positive health outcomes.

Paragraph 1 Summary

Unhealthy food is linked to disease incidence, so government interventions on food prices could help prevent disease.

[P2] Utilizing national data from 2012 regarding food consumption, health, and economic status, Peñalvo et al. concluded that such price adjustments would help to prevent deaths related to cardiometabolic diseases. According to their analysis, increasing the prices of unhealthy foods such as processed meats and sugary sodas by 10%, while reducing the prices of healthy foods such as fruit and vegetables by 10%, would prevent an estimated 3.4% of yearly deaths in the U.S. Changing prices by 30% would have an even stronger effect, preventing an estimated 9.2% of yearly deaths. This data comports with that found in other countries, such as "previous modeling studies in South Africa and India, where a 20% SSB [sugar-sweetened beverage] tax was estimated to reduce diabetes prevalence by 4% over 20 years." The effects of price adjustments would be most pronounced on persons of lower socioeconomic status, as the researchers "found an overall 18.2% higher price-responsiveness for low versus high SES [socioeconomic status] groups."

Paragraph 2 Summary

Multiple data sets support the previous hypothesis: governments can prevent disease by making healthy food more affordable and unhealthy food less affordable.

[P3] [1] This differential effect based on socioeconomic status contributes to concerns about such interventions, however. In *Harvard Public Health Review*, Kates and Hayward ask: "Well intentioned though they may be, at what point do these taxes overstep government influence on an individual's right to autonomy in decision-making? On whom does the increased financial burden of this taxation fall?" They note that taxes on sugar-sweetened beverages, for instance, "are likely to have a greater impact on low-income individuals…because individuals in those settings are more likely to be beholden to cost when making decisions about food."

[1] Topic Shift

Up until this point, the passage has focused on whether adjusting food prices is *effective* in preventing disease. But now the author is discussing whether such interventions *ought* to be implemented.

Paragraph 3 Summary

Some worry that government price interventions may violate consumers' autonomy or unfairly burden the poor.

[P4] [2] However, "well intentioned though they may be," the worries that Kates and Hayward express are to some extent misguided. In particular, the idea that taxing unhealthy foods would burden those least able to afford it misses the point. Although the increased taxes would affect anyone who continued to purchase the items despite the higher prices, the goal of raising prices on unhealthy foods is precisely to dissuade people from buying them. As Kates and Hayward themselves remark, "Those in low-income environments may also be the largest consumers of obesogenic foods and therefore most likely to benefit from such a lifestyle change indirectly posed by SSB taxes." As the goal of the taxes is to promote those lifestyle changes, the financial burden objection is a non-starter.

[P5] Given this recognition, the question regarding autonomy constitutes a more substantial issue. [3] Nevertheless, that concern also rests on a dubious assumption, as people's autonomy is not necessarily respected in the current situation either. The fact that those of lower socioeconomic status are more likely to have poorer diets suggests that such persons' food choices are the result of financial constraint, not fully autonomous, rational deliberation. Hence, by making healthy foods more affordable relative to unhealthy ones, government intervention might actually *facilitate* autonomous choices rather than hindering them.

[P6] [4] On the other hand, suppose that the disproportionate consumption of unhealthy foods—and associated higher incidence of disease—among certain groups is not the result of financial hardship but rather the result of those persons' perceived self-interest. If so, that would suggest that members of these groups are being encouraged to persist in harmful dietary habits for the sake of corporate profits. In that case, violating autonomy for the sake of health may be permissible, as that practice would be morally preferable to the present system of corporate exploitation.

Food Costs and Disease ©UWorld

---

[2] Addressing Opposing View

The author argues against the objection that price interventions might unethically burden low-income individuals.

Paragraph 4 Summary

The author argues that Kates and Hayward's concern about burdening the poor misses the point of the price interventions.

[3] Addressing Opposing View

The author argues against the objection that price interventions might unethically violate people's autonomy.

Paragraph 5 Summary

The author agrees that autonomy is a legitimate concern, but suggests that price interventions might not violate it after all.

[4] Connected Ideas

The author previously argued that price interventions are justified if poor diet is caused by financial constraint. Now they consider whether those interventions might still be justified even if poor diet is *not* caused by financial constraint.

Paragraph 6 Summary

The author argues that autonomy is violated whether or not there are price interventions, so it's better to violate it for improved health than for corporate profits.

Passage G: Food Costs and Disease

Content Annotations

Passage Observations: 8 Annotations
Potential Pitfalls: 3 Annotations

[P1] Because frequent consumption of unhealthy foods is strongly linked with cardiometabolic diseases, one way for governments to combat those afflictions may be to modify the eating habits of the general public. [1] Applying economic incentives or disincentives to various types of foods could potentially alter people's diets, leading to more positive health outcomes.

[1] **Focus**

The passage seems to be exploring two relationships: how people's diets affect their incidence of disease; and how government price interventions affect people's diets.

[P2] Utilizing national data from 2012 regarding food consumption, health, and economic status, [2] Peñalvo et al. concluded that such price adjustments would help to prevent deaths related to cardiometabolic diseases. According to their analysis, increasing the prices of unhealthy foods such as processed meats and sugary sodas by 10%, while reducing the prices of healthy foods such as fruit and vegetables by 10%, would prevent an estimated 3.4% of yearly deaths in the U.S. Changing prices by 30% would have an even stronger effect, preventing an estimated 9.2% of yearly deaths. [3] This data comports with that found in other countries, such as "previous modeling studies in South Africa and India, where a 20% SSB [sugar-sweetened beverage] tax was estimated to reduce diabetes prevalence by 4% over 20 years." [4] The effects of price adjustments would be most pronounced on persons of lower socioeconomic status, as the researchers "found an overall 18.2% higher price-responsiveness for low versus high SES [socioeconomic status] groups."

[2] **Outside Source**

Peñalvo et al.'s research supports the claim that adjustments to food prices have a positive effect on disease prevention.

[3] **Outside Source**

Research from other countries supports the same relationship between food prices and disease prevention.

[4] **Likely Question Topic**

Socioeconomic status adds another variable to the relationship between price adjustments and health outcomes. Thus, it is likely to play a role in questions about price interventions.

[P3] This differential effect based on socioeconomic status contributes to concerns about such interventions, however. [5] In *Harvard Public Health Review*, Kates and Hayward ask: "Well intentioned though they may be, at what point do these taxes overstep government influence on an individual's right to autonomy in decision-making? On whom does the increased financial burden of this taxation fall?" [1] They note that taxes on sugar-sweetened beverages, for instance, "are likely to have a greater impact on low-income individuals...because individuals

[5] **Outside Source**

Kates and Hayward are cited to raise ethical concerns about food price interventions.

[1] **Potential Pitfall: Missed Distinction**

Kates and Hayward are not disputing the data about price interventions, only whether such interventions are ethical. Thus, thinking of them as just "disagreeing with price interventions" could make a wrong answer appear correct.

in those settings are more likely to be beholden to cost when making decisions about food."

[P4] However, "well intentioned though they may be," the worries that Kates and Hayward express are to some extent misguided. In particular, the idea that taxing unhealthy foods would burden those least able to afford it misses the point. Although the increased taxes would affect anyone who continued to purchase the items despite the higher prices, the goal of raising prices on unhealthy foods is precisely to dissuade people from buying them. [2] As Kates and Hayward themselves remark, "Those in low-income environments may also be the largest consumers of obesogenic foods and therefore most likely to benefit from such a lifestyle change indirectly posed by SSB taxes." [6] As the goal of the taxes is to promote those lifestyle changes, the financial burden objection is a non-starter.

[P5] [7] Given this recognition, the question regarding autonomy constitutes a more substantial issue. Nevertheless, that concern also rests on a dubious assumption, as people's autonomy is not necessarily respected in the current situation either. The fact that those of lower socioeconomic status are more likely to have poorer diets suggests that such persons' food choices are the result of financial constraint, not fully autonomous, rational deliberation. [3] Hence, by making healthy foods more affordable relative to unhealthy ones, government intervention might actually *facilitate* autonomous choices rather than hindering them.

[P6] On the other hand, suppose that the disproportionate consumption of unhealthy foods—and associated higher incidence of disease—among certain groups is not the result of financial hardship but rather the result of those persons' perceived self-interest. If so, that would suggest that members of these groups are being encouraged to persist in harmful dietary habits for the sake of corporate profits. [8] In that case, violating autonomy for the sake of health may be permissible, as that practice would be morally preferable to the present system of corporate exploitation.

Food Costs and Disease ©UWorld

---

[2] **Potential Pitfall: Missed Distinction**

There is no disagreement about the facts of price interventions, only their ethical implications. Thus, thinking of the author as just "disagreeing with Kates and Hayward" could make a wrong answer appear correct.

[6] **Author's View**

According to the author, price interventions do disproportionately impact the poor, but this is an intended feature of the interventions rather than a drawback.

[7] **Author's View**

The author believes that concerns about autonomy are significant, unlike the previous objection to price interventions (about the impact on low-income people).

[3] **Potential Pitfall: Missed Distinction**

Note that the author is not arguing that it is okay to violate autonomy here. Rather, they state that price interventions might increase autonomy instead. Overlooking this distinction could lead to mistakes about the author's position.

[8] **Focus**

The author agrees that autonomy is valuable but argues that violating it for health is more ethical than allowing corporate exploitation. Especially given the second half of the passage, this conclusion suggests that the author's main purpose is to defend the use of government price interventions to improve public health.

Passage H: For Whom the Bell Toils

This passage is associated with 5 questions analyzed in the UWorld MCAT CARS book.

- Subskill 1c. Direct Passage Claims Question 3
- Subskill 1f. Further Implications of Passage Claims Question 3
- Subskill 2f. Drawing Additional Inferences Question 1
- Subskill 3c. New Claim Support or Challenge Question 2
- Subskill 3f. Additional Conclusions From New Information Question 1

Passage H: For Whom the Bell Toils

Structural Annotations

Passage Structure: 6 Annotations

[P1] In nineteenth-century America, most people dismissed the notion that someone might assassinate the president. The presumption was based not only on ethics but practicality: a president's term is inherently limited, and an unpopular one could be voted out of office. Therefore, it was reasoned, there would be no need to consider removal through violence. This belief persisted even after the shocking murder of Abraham Lincoln in 1865, which was viewed as an aberration. Thus it was that on July 2, 1881, Charles Guiteau could simply walk up to President James A. Garfield and shoot him in broad daylight. As Richard Menke portrays events, "Guiteau was in fact a madman who had come to identify with a disgruntled wing of the Republican Party after his deranged fantasies of winning a post from the new administration had come to nothing." Believing that God had told him to kill the president, Guiteau thought this act would garner fame for his religious ideas and thereby help to usher in the Apocalypse.

Paragraph 1 Summary

People used to believe that no one would consider assassinating a president, so Charles Guiteau could easily shoot President Garfield in 1881. Guiteau was motivated both by his political frustration and his radical religious views.

[P2] [1] In an interesting parallel, Garfield had felt a sense of divine purpose for his own life after surviving a near-drowning as a young man. Unlike Guiteau's fanatical ravings, however, Garfield's vision worked to the betterment of himself and the world. Candice Millard describes his ascent from extreme poverty to incredible excellence in college, where "by his second year...they made him a professor of literature, mathematics, and ancient languages." Garfield would go on to join the Union Army, where he attained the rank of major general and argued that black soldiers should receive the same pay as their white compatriots. While serving in the war he was nominated for the House of Representatives but accepted the seat only after President Lincoln declared that the country had more need of him as a congressman than as a general. The reluctant politician would later himself become president under similar circumstances, after multiple factions of a deadlocked Republican convention unexpectedly nominated him instead of their original candidates in 1880. An honest man who opposed corruption within the party, Garfield strove both to heal the

[1] **Topic Shift**

Here the passage switches from discussing the assassination to providing biographical details about Garfield.

Paragraph 2 Summary

The author contrasts Guiteau's delusion with Garfield's aspiration. In addition to extreme academic achievement, Garfield served in the Union Army and argued for more racial fairness.

fractures of the Civil War and to uphold the aims for which it was fought, until "the equal sunlight of liberty shall shine upon every man, black or white, in the Union."

[P3] [2] Although Guiteau's bullet would ultimately dim this light for Garfield, the president actually survived the initial attack and for a time appeared headed for recovery. Tragically, however, the hubris shown by his main physician, Dr. Willard Bliss, would lead instead to weeks of prolonged suffering. None of the doctors who examined Garfield were able to locate the bullet, and its lingering presence—along with the unwashed hands of the doctors who probed for it—led to an infection. As the president's condition worsened, inventor Alexander Graham Bell attempted to adapt his patented telephone technology to locate foreign metal in the human body. Inspired by speculation that the bullet's electromagnetic properties might be detectable, Bell used his newly developed "Induction Balance" device to listen for the sounds of electrical interference he hoped would isolate the site of the bullet.

[P4] Unfortunately, Bell's searches were unsuccessful. Like Garfield's doctors, he had been looking for the bullet in the wrong area. Menke asserts that Bell's efforts "would probably have fallen short" regardless. However, other historians suggest that Dr. Bliss, unwilling to consider challenges to his original assessment, prevented Bell from more thoroughly searching the president's body. Certainly, Bliss ignored the advice and protestations of other physicians, even as Garfield continued to decline. With death imminent, Garfield asked to be taken to his seaside cottage, where he died on the 19th of September.

For Whom the Bell Toils ©UWorld

He became both a congressman and eventually president almost by accident, using those offices to fight for honesty and equality.

[2] **Topic Shift**

The passage now returns to the original topic of Garfield's assassination.

Paragraph 3 Summary

Garfield seemed like he might recover, but Dr. Bliss' pride caused him to keep suffering instead. Because doctors were unable to locate the bullet, Alexander Graham Bell tried to make a device to find it.

Paragraph 4 Summary

Bell failed to find the bullet, possibly in part due to Dr. Bliss' interference. Dr. Bliss also refused to listen to other doctors despite Garfield's continued decline, and Garfield died under his care.

Passage H: For Whom the Bell Toils

Content Annotations

Passage Observations: 11 Annotations
Potential Pitfalls: 1 Annotation

[P1] [1] In nineteenth-century America, most people dismissed the notion that someone might assassinate the president. The presumption was based not only on ethics but practicality: a president's term is inherently limited, and an unpopular one could be voted out of office. Therefore, it was reasoned, there would be no need to consider removal through violence. This belief persisted even after the shocking murder of Abraham Lincoln in 1865, which was viewed as an aberration. [2] Thus it was that on July 2, 1881, Charles Guiteau could simply walk up to President James A. Garfield and shoot him in broad daylight. [3] As Richard Menke portrays events, "Guiteau was in fact a madman who had come to identify with a disgruntled wing of the Republican Party after his deranged fantasies of winning a post from the new administration had come to nothing." Believing that God had told him to kill the president, Guiteau thought this act would garner fame for his religious ideas and thereby help to usher in the Apocalypse.

[P2] [4] In an interesting parallel, Garfield had felt a sense of divine purpose for his own life after surviving a near-drowning as a young man. Unlike Guiteau's fanatical ravings, however, Garfield's vision worked to the betterment of himself and the world. [5] Candice Millard describes his ascent from extreme poverty to incredible excellence in college, where "by his second year…they made him a professor of literature, mathematics, and ancient languages." Garfield would go on to join the Union Army, where he attained the rank of major general and argued that black soldiers should receive the same pay as their white compatriots. [6] While serving in the war he was nominated for the House of Representatives but accepted the seat only after President Lincoln declared that the country had more need of him as a congressman than as a general. The reluctant politician would later himself become president under similar circumstances, after multiple factions of a deadlocked Republican convention unexpectedly nominated him instead of their original candidates in 1880. An honest man who opposed corruption within the party, [7] Garfield strove

[1] **Claim**

Part of the historical context surrounding Garfield's assassination was the widespread belief that no one would consider such a thing.

[2] **Focus**

Based on this statement and how the preceding information leads up to it, the passage will likely center on describing the circumstances of Garfield's assassination.

[3] **Outside Source**

The author cites Menke's words to explain Guiteau's motivations: political revenge and religious delusions.

[4] **Author's View**

The author depicts Garfield and Guiteau as mirror images of each other: one embodied intelligence and moral courage, the other lunacy and destructive violence.

[5] **Outside Source**

Millard's words illustrate Garfield's extraordinary mind—he was "promoted" from college student to professor.

[6] **Claim**

Garfield became both a congressman and eventually the president despite not intentionally running for either office.

[7] **Likely Question Topic**

This is a dense paragraph in which the author highlights several unusual aspects of Garfield's

both to heal the fractures of the Civil War and to uphold the aims for which it was fought, until "the equal sunlight of liberty shall shine upon every man, black or white, in the Union."

[P3] Although Guiteau's bullet would ultimately dim this light for Garfield, [8] the president actually survived the initial attack and for a time appeared headed for recovery. Tragically, however, the hubris shown by his main physician, Dr. Willard Bliss, would lead instead to weeks of prolonged suffering. [1] None of the doctors who examined Garfield were able to locate the bullet, and its lingering presence—along with the unwashed hands of the doctors who probed for it—led to an infection. As the president's condition worsened, [9] inventor Alexander Graham Bell attempted to adapt his patented telephone technology to locate foreign metal in the human body. Inspired by speculation that the bullet's electromagnetic properties might be detectable, Bell used his newly developed "Induction Balance" device to listen for the sounds of electrical interference he hoped would isolate the site of the bullet.

[P4] Unfortunately, Bell's searches were unsuccessful. Like Garfield's doctors, he had been looking for the bullet in the wrong area. [10] Menke asserts that Bell's efforts "would probably have fallen short" regardless. However, other historians suggest that Dr. Bliss, unwilling to consider challenges to his original assessment, prevented Bell from more thoroughly searching the president's body. [11] Certainly, Bliss ignored the advice and protestations of other physicians, even as Garfield continued to decline. With death imminent, Garfield asked to be taken to his seaside cottage, where he died on the 19th of September.

For Whom the Bell Toils ©UWorld

life. It would be reasonable to expect a question about Garfield's unconventional political career or his impressive accomplishments and character.

[8] Author's View

The author implies that Dr. Bliss was partially to blame for Garfield's suffering and death.

[1] Potential Pitfall: Misreading

Since the author just said Dr. Bliss' hubris caused Garfield to suffer, it would be natural to expect this next sentence to provide details. But it doesn't actually tell us anything more about Dr. Bliss or his hubris; it explains only why Garfield's wound became infected. Thus, the proximity of the sentences and our natural expectations could lead to mistakes about the point of this claim.

[9] Claim

Alexander Graham Bell attempted to develop a device that could locate the bullet and thereby help save Garfield.

[10] Outside Source

Based on the disagreement between Menke and other historians, it is unclear whether Bell might have found the bullet without Bliss' interference.

[11] Elaboration

The previous paragraph stated that Dr. Bliss' hubris caused Garfield to keep suffering instead of recovering. However, it is only now that the author specifies: Bliss might have kept Bell from finding the bullet, and he definitely refused to listen to other doctors' dissenting opinions even though Garfield was dying.

Passage I: Meaning: Readers or Authors?

This passage is associated with 5 questions analyzed in the UWorld MCAT CARS book.

- Subskill 1c. Direct Passage Claims Question 5
- Subskill 2f. Drawing Additional Inferences Question 3
- Subskill 3d. External Scenario Support or Challenge Question 4
- Subskill 3e. Applying Passage Perspectives Question 2
- Subskill 3g. Identifying Analogies Question 2

Passage I: Meaning: Readers or Authors?

Structural Annotations

Passage Structure: 9 Annotations

[P1] Of late it has become popular among linguists and literary theorists to assert that a work's *meaning* depends upon the individual reader. It is readers, we are told, not authors, who create meaning, by interacting with a text rather than simply receiving it. Thus, a reader transcends the aims of the author, producing their own reading of the text. Indeed, on this line of thinking, even to speak of "the" text is to commit a conceptual error; every text is in fact many texts, a plurality of interpretations that resist comparative evaluation. This view is nonsense. That many otherwise sensible scholars should be attracted to it can perhaps be readily explained, but we should first delineate why the theory goes so far astray....

Paragraph 1 Summary

The passage author strongly opposes the view of literary interpretation that a work's meaning depends on readers rather than authors, and that those readers' different interpretations resist evaluation. The passage will go on to explain why that view should be considered nonsense.

[P2] The absurdity of the view can be demonstrated by a practical analogy. Suppose Smith is conveying his ideas to Jones in conversation (the particular topic is of no consequence). Afterward, we discover that the men differ in their accounts of what Smith had expressed. At this point, Jones may decide that he misunderstood Smith, or perhaps that Smith was unclear. A more complex supposition might be that Smith misused some key term, so his words did not fully match his intentions. Any of these possibilities would reasonably describe why Jones and Smith possessed different opinions about what Smith had said.

Paragraph 2 Summary

The passage author describes a situation where one man misunderstands another, as part of an analogy meant to show that the reader-based view of meaning is absurd.

[P3] [3] What Jones may *not* justifiably conclude is that his own interpretation is what Smith *really meant*. He may not, in effect, say: "Yes, I admit that Smith honestly claims to have been saying something different, but I have formed my own equally correct understanding." Someone who made such an assertion would be suspected of making a joke; if he proved to be serious, we could only conclude that he was deeply confused or else being deliberately quarrelsome. For in questions about what Smith meant, it is surely Smith whose answer must be accepted.... [T]his is not a matter of *agreeing* with a speaker; Jones might judge Smith's ideas to be wrong, unfounded, etc. But whether Smith's ideas are right or wrong is a different

[3] Connected Ideas

The passage author began describing the analogy in the previous paragraph, but it is here that the argument about it actually starts.

Paragraph 3 Summary

In the passage author's analogy, a listener knows he misunderstood what a speaker meant but still claims his own interpretation is "equally correct." The listener's position is portrayed as

matter from what those ideas *are*. On that count, Smith must be the authority.

[P4] ⁵ However, this observation is in no way changed if Smith's ideas are written rather than spoken—sent by letter, for instance. Regardless of any interpretation Jones may produce, the letter's true meaning is whatever Smith intended to convey. Likewise it is, then, with a book, poem, or whatsoever object of literature a scholar (or ordinary reader) encounters. The writing down of ideas does not magically imbue them with malleability or render their content amorphous. From the loftiest tomes of Shakespeare or Milton to the lowliest of yellowed paperbacks, authors produce works with a particular message in mind. ⁶ It is readers' task to discern that message, not to superimpose their own volitional perspectives.

[P5] To think otherwise is to undermine the foundation of literary scholarship. For what is the purpose of such scholarship, if not to seek understanding of an author's creation? One examines the text, taking note of style, historical context, allusions to other works, and other factors, in addition, of course, to the surface sense of the words themselves. ⁷ If such an enterprise is to be reasonable, it must presume the existence of standards for success: accuracy and inaccuracy, depth or shallowness of analysis, grounds for preferring one interpretation to another. Different readers may come to different conclusions about a text, it is true. But to excise authorial intent from the evaluation of those conclusions does a disservice both to individual works and to literary study as a discipline.

Meaning: Readers or Authors? ©UWorld

ridiculous, since it is clearly the speaker himself who knows what he really meant.

⁵ Connected Ideas

The passage author's analogy and associated argument extend from the second to the fourth paragraphs, forming the bulk of the passage.

⁶ Connected Ideas

This claim refers back to the first paragraph, again rejecting the idea that the reader "transcends the aims of the author."

Paragraph 4 Summary

The passage author completes their argument by analogy: Just as it is absurd to think a listener determines the meaning of what a speaker says, it is also absurd to think a reader determines the meaning of what an author writes.

⁷ Connected Ideas

The author again refers back to the first paragraph, rejecting the idea that different interpretations "resist comparative evaluation."

Paragraph 5 Summary

The passage author concludes that literary meaning is determined by an author's intention. To accept the alternative, reader-based theory of meaning is a misguided and harmful approach to literary scholarship.

Passage I: Meaning: Readers or Authors?

Content Annotations

Passage Observations: 9 Annotations
Potential Pitfalls: 1 Annotation

[P1] [1] Of late it has become popular among linguists and literary theorists to assert that a work's *meaning* depends upon the individual reader. It is readers, we are told, not authors, who create meaning, by interacting with a text rather than simply receiving it. Thus, a reader transcends the aims of the author, producing their own reading of the text. Indeed, on this line of thinking, even to speak of "the" text is to commit a conceptual error; every text is in fact many texts, a plurality of interpretations that resist comparative evaluation. [2] This view is nonsense. That many otherwise sensible scholars should be attracted to it can perhaps be readily explained, but [3] we should first delineate why the theory goes so far astray....

[P2] [4] The absurdity of the view can be demonstrated by a practical analogy. Suppose Smith is conveying his ideas to Jones in conversation (the particular topic is of no consequence). Afterward, we discover that the men differ in their accounts of what Smith had expressed. At this point, Jones may decide that he misunderstood Smith, or perhaps that Smith was unclear. A more complex supposition might be that Smith misused some key term, so his words did not fully match his intentions. Any of these possibilities would reasonably describe why Jones and Smith possessed different opinions about what Smith had said.

[P3] What Jones may *not* justifiably conclude is that his own interpretation is what Smith *really meant*. He may not, in effect, say: "Yes, I admit that Smith honestly claims to have been saying something different, but I have formed my own equally correct understanding." Someone who made such an assertion would be suspected of making a joke; if he proved to be serious, we could only conclude that he was deeply confused or else being deliberately quarrelsome. [5] For in questions about what Smith meant, it is surely Smith whose answer must be accepted.... [6] [T]his is not a matter of *agreeing* with a speaker; Jones might judge Smith's ideas to be wrong,

[1] Claim

According to one popular view, the meaning of a literary work is highly subjective and varies based on each individual reader.

[2] Author's View

The passage author not only disagrees with the view being described but expresses a strongly negative attitude toward it.

[3] Focus

The purpose of the passage seems to be to argue against the reader-based view of meaning just described.

[4] Author's View

The passage author again criticizes the view being discussed, referring to its "absurdity."

[5] Author's View

The passage author seems to take this point for granted (it is "surely" correct). Thus, this claim—that it is the speaker, not the listener, who knows what the speaker really meant—is likely a key point in the passage author's overall argument.

[6] Elaboration

The passage author distinguishes between knowing what a speaker means and whether

unfounded, etc. But whether Smith's ideas are right or wrong is a different matter from what those ideas *are*. On that count, Smith must be the authority.

[P4] However, this observation is in no way changed if Smith's ideas are written rather than spoken—sent by letter, for instance. Regardless of any interpretation Jones may produce, the letter's true meaning is whatever Smith intended to convey. [7] Likewise it is, then, with a book, poem, or whatsoever object of literature a scholar (or ordinary reader) encounters. [8] The writing down of ideas does not magically imbue them with malleability or render their content amorphous. From the loftiest tomes of Shakespeare or Milton to the lowliest of yellowed paperbacks, authors produce works with a particular message in mind. It is readers' task to discern that message, not to superimpose their own volitional perspectives.

[P5] To think otherwise is to undermine the foundation of literary scholarship. For what is the purpose of such scholarship, if not to seek understanding of an author's creation? [9] One examines the text, taking note of style, historical context, allusions to other works, and other factors, in addition, of course, to the surface sense of the words themselves. If such an enterprise is to be reasonable, it must presume the existence of standards for success: accuracy and inaccuracy, depth or shallowness of analysis, grounds for preferring one interpretation to another. [1] Different readers may come to different conclusions about a text, it is true. But to excise authorial intent from the evaluation of those conclusions does a disservice both to individual works and to literary study as a discipline.

Meaning: Readers or Authors? ©UWorld

---

that speaker is correct. People always know what their own ideas are, even if those ideas are misguided or wrong.

[7] Elaboration

According to the passage author, their conclusion about meaning applies to all forms of writing and all types of readers.

[8] Author's View

The use of the word "magically" further characterizes the reader-based view of meaning as nonsensical.

[9] Author's View

Here the passage author describes how they think literary interpretation ought to work (rather than the reader-based view of meaning they have been arguing against).

[1] Potential Pitfall: Missed Distinction

Note that the passage author never disputes the idea that interpretations of a work may differ; they simply reject the idea that such interpretations are equally correct. Missing this point could lead to mistakes about the passage author's position.

Passage J: The Inkblots

This passage is associated with 5 questions analyzed in the UWorld MCAT CARS book.

- Subskill 1d. Implicit Claims or Assumptions Question 3
- Subskill 1e. Identifying Passage Perspectives Question 3
- Subskill 2c. Extent of Passage Evidence Question 2
- Subskill 2e. Determining Passage Perspectives Question 2
- Subskill 3a. Exemplar Scenario for Passage Claims Question 2

Passage J: The Inkblots

Structural Annotations

Passage Structure: 7 Annotations

[P1] For almost 100 years now, the psychological evaluation known as the Rorschach Inkblot Test has engendered much controversy, including skepticism about its value, questions about its scoring, and, especially, criticism of its interpretive methods as too subjective. Thus, the Rorschach test, which emerged from the same early twentieth-century zeitgeist that produced Einstein's physics, Freudian psychoanalysis, and abstract art, seems one of modernity's most misbegotten children. Destined never to be completely accepted or discredited, the test remains a perennial outlier in its field. Nevertheless, the inkblots' mystery and aesthetic appeal have caused them to be indelibly printed on our cultural fabric.

Paragraph 1 Summary

The Rorschach test has long been controversial, especially because it is criticized as too subjective. Nevertheless, it continues to have great cultural impact.

[P2] The now iconic inkblots were introduced to the world by Swiss psychiatrist Hermann Rorschach in his 1921 book *Psychodiagnostics*. As both director of the Herisau Asylum in Switzerland and an amateur artist, Rorschach was uniquely positioned to wed the new practice of psychoanalysis to the budding phenomenon of abstract art. For instance, reading Freud's work on dream symbolism prompted him to recall his childhood passion for a game based on inkblot art called *Klecksographie*. He was also cognizant that in a recently published dissertation, his colleague Szymon Hens had used inkblots to try to probe the imagination of research subjects; moreover, a few years earlier, the French psychologist and father of intelligence testing Alfred Binet had used them to measure creativity.

Paragraph 2 Summary

Rorschach had many influences in creating his test: his love of art, Freud's dream symbolism, and other scientists' use of inkblots in research

[P3] Motivated by these developments, the Herisau director decided to revisit that childhood pastime that had awakened his curiosity about how visual information is processed. In particular, he wondered why different people saw different things in the same image. Traditionally, psychoanalysts had relied on language for insights; however, as biographer Damion Searls reports, Rorschach's theories would exemplify the principle that "who we are is a matter less of what we say than of what we see." Indeed, through a process of perception termed pareidolia, the mind projects meaning onto images,

Paragraph 3 Summary

Most psychoanalysis had been based on language, but the inkblot test was instead based

detecting in them familiar objects or shapes. Consequently, what a person sees in an image reveals more about that person than about the image itself.

[P4] Rorschach experimented with countless inkblots, eventually selecting ten—five black on white, two also featuring some red, and three pastel-colored—to use with research subjects. For these perfectly symmetrical images—each of which he was said to have "meticulously designed to be as ambiguous and 'conflicted' as possible"—the primary question was always "What do *you* see?" Rorschach was especially careful to note how much attention individuals paid to various components of each inkblot (such as form, color, and a sense of movement) and whether they concentrated on details or the whole image. Having observed that his patients with schizophrenia gave distinctly different responses from the control group, Rorschach envisioned his experiment as a diagnostic tool for the disease. Nevertheless, he resisted the notion that its results could be used to assess personality. In fact, until his untimely death from a ruptured appendix in 1922, Rorschach referred to his project as an "interpretive form experiment" rather than a test. [1] Ironically, however, by the 1960s, the Rorschach Inkblot Test was known chiefly as a personality assessment and had become the most frequently administered projective personality test in the US.

[P5] [2] Rorschach's test has survived nearly incessant scrutiny, including a 2013 comprehensive study of all Rorschach test data and repeated revisions to its scoring, yet doubts about its validity and reliability persist. Much like the inkblots themselves—which tantalize us with the possibility of divulging the secrets of who we are and how we see the world—the test has (for better or worse) defied attempts to fix its meaning. Thus, what has been called "the twentieth century's most visionary synthesis of art and science" stands tempered by harsh criticism.

The Inkblots ©UWorld

on visual interpretation. Each person's mind may impart a different meaning to the same image.

[1] Connected Ideas

The test became primarily used in a way that Rorschach never intended. Thus, this fact may relate to the inkblots' controversial aspects as mentioned in Paragraph 1.

Paragraph 4 Summary

The inkblots are ambiguous and elicit varying interpretations, and Rorschach envisioned them as helping to diagnose schizophrenia. Ironically, however, they became primarily used as a general personality assessment, something that Rorschach himself resisted.

[2] Topic Shift

Although most of the passage has discussed the Rorschach test's development, the author now returns to the controversy mentioned in the first paragraph.

Paragraph 5 Summary

The Rorschach test remains controversial, continuing both to hold great appeal and to receive strong criticism.

Passage J: The Inkblots

Content Annotations

Passage Observations: 11 Annotations
Potential Pitfalls: 1 Annotations

[P1] For almost 100 years now, the psychological evaluation known as [1] the Rorschach Inkblot Test has engendered much controversy, including skepticism about its value, questions about its scoring, and, especially, criticism of its interpretive methods as too subjective. Thus, the Rorschach test, which emerged from the same early twentieth-century zeitgeist that produced Einstein's physics, Freudian psychoanalysis, and abstract art, seems one of modernity's most misbegotten children. [2] Destined never to be completely accepted or discredited, the test remains a perennial outlier in its field. Nevertheless, the inkblots' mystery and aesthetic appeal have caused them to be indelibly printed on our cultural fabric.

[P2] The now iconic inkblots were introduced to the world by Swiss psychiatrist Hermann Rorschach in his 1921 book *Psychodiagnostics*. As both director of the Herisau Asylum in Switzerland and an amateur artist, [3] Rorschach was uniquely positioned to wed the new practice of psychoanalysis to the budding phenomenon of abstract art. For instance, [4] reading Freud's work on dream symbolism prompted him to recall his childhood passion for a game based on inkblot art called *Klecksographie*. He was also cognizant that in a recently published dissertation, his colleague Szymon Hens had used inkblots to try to probe the imagination of research subjects; moreover, a few years earlier, the French psychologist and father of intelligence testing Alfred Binet had used them to measure creativity.

[P3] Motivated by these developments, the Herisau director decided to revisit that childhood pastime that had awakened his curiosity about how visual information is processed. In particular, he wondered why different people saw different things in the same image. [5] Traditionally, psychoanalysts had relied on language for insights; however, as biographer Damion Searls reports, Rorschach's theories would exemplify the principle that "who we are is a matter less of what we say than of what we see." [6] Indeed, through a process of perception

---

[1] Claim

The Rorschach test has been considered controversial for several reasons, especially its subjective nature.

[2] Focus

The introduction seemed to be emphasizing the controversy around the Rorschach test, but it ends by talking about the inkblots' effect on our culture. Thus, the passage is likely a more descriptive account of the inkblots, rather than an argument about their controversial aspects.

[3] Claim

The Rorschach test combines psychoanalysis and art, which was one of Rorschach's interests.

[4] Likely Question Topic

This list of specific individuals and works could easily generate questions about Rorschach's inspirations and the scholarship that led up to the test's development.

[5] Outside Source

Searls is quoted to explain how Rorschach's approach was atypical. While most psychoanalysis had been based on language, the Rorschach test instead focused on the interpretation of visual stimuli.

[6] Likely Question Topic

By connecting the Rorschach test to a specific psychological phenomenon, this information

termed pareidolia, the mind projects meaning onto images, detecting in them familiar objects or shapes. Consequently, what a person sees in an image reveals more about that person than about the image itself.

[P4] [7] Rorschach experimented with countless inkblots, eventually selecting ten—five black on white, two also featuring some red, and three pastel-colored—to use with research subjects. For these perfectly symmetrical images—each of which he was said to have [8] "meticulously designed to be as ambiguous and 'conflicted' as possible"—the primary question was always "What do *you* see?" Rorschach was especially careful to note how much attention individuals paid to various components of each inkblot (such as form, color, and a sense of movement) and whether they concentrated on details or the whole image. Having observed that his [9] patients with schizophrenia gave distinctly different responses from the control group, Rorschach envisioned his experiment as a diagnostic tool for the disease. Nevertheless, he resisted the notion that its results could be used to assess personality. In fact, until his untimely death from a ruptured appendix in 1922, Rorschach referred to his project as an "interpretive form experiment" rather than a test. [1] Ironically, however, by the 1960s, the Rorschach Inkblot Test was known chiefly as a personality assessment and had become the most frequently administered projective personality test in the US.

[P5] Rorschach's test has survived nearly incessant scrutiny, including a 2013 comprehensive study of all Rorschach test data and repeated revisions to its scoring, yet doubts about its validity and reliability persist. [10] Much like the inkblots themselves—which tantalize us with the possibility of divulging the secrets of who we are and how we see the world—the test has (for better or worse) defied attempts to fix its meaning. [11] Thus, what has been called "the twentieth century's most visionary synthesis of art and science" stands tempered by harsh criticism.

The Inkblots ©UWorld

---

could generate questions about pareidolia, the functioning of the Rorschach test, or both.

[7] Elaboration

This is the first (and only) physical description of the inkblots in the passage.

[8] Elaboration

The test functions through the inkblots' ambiguity: people project their own meaning onto part or all of a given image.

[9] Claim

Rorschach thought his inkblots could help diagnose schizophrenia, but he specifically resisted the idea that they represented a test for assessing personality. The term "resisted" implies that others did believe the inkblots should be used in that way, revealing a conflict over their purpose.

[1] Potential Pitfall: Details

This paragraph presents a fair amount of overlapping information, including distinctions concerning how the Rorschach test functioned, how Rorschach envisioned its purpose, and how it has mostly been used. It would be easy to mix up some of this information when approaching questions.

[10] Author's View

Just as the inkblots are known for their ambiguity, the author characterizes the Rorschach test as ambiguous overall.

[11] Focus

As in the first paragraph, the passage does not argue for or against the validity or reliability of the Rorschach test, but simply affirms that such criticism persists. Thus, the purpose of the passage as a whole seems to be to describe the test's development and the controversy that surrounds it.

Passage K: Lengthening the School Day

This passage is associated with 7 questions analyzed in the UWorld MCAT CARS book.

- Subskill 1f. Further Implications of Passage Claims Question 1
- Subskill 2a. Logical Relationships Within Passage Question 3
- Subskill 2a. Logical Relationships Within Passage Question 4
- Subskill 2c. Determining Passage Perspectives Question 3
- Subskill 3c. External Scenario Support or Challenge Question 1
- Subskill 3d. Additional Conclusions From New Information Question 3
- Subskill 3e. Applying Passage Perspectives Question 1

Passage K: Lengthening the School Day

Structural Annotations

Passage Structure: 8 Annotations

[P1] There may be reasons to reject the idea of lengthening the school day. None of them, however, are *good* reasons. Rather, the supposed demerits of such a proposal fall easily in the face of its numerous financial and social benefits for families.

Paragraph 1 Summary

The author will argue that lengthening the school day has many benefits, and that any supposed drawbacks can be dismissed.

[P2] The greatest of these benefits lies in reducing the need for childcare. It is a curious fact of American life that the adult's work schedule and the child's school schedule are misaligned. Children rise with the sun to head to classes, only to be sent home again hours before parents return from their jobs. In a society where, more often than not, both parents work, this discordance creates the need for an expensive arrangement to fill the gap in families' routines. For instance, studies show that in 2016, childcare costs accounted for 9.5 to 17.5 percent of median family income, depending on the state. Today, 40 percent of families nationwide spend over 15 percent of their income on childcare. Transportation to and from care sites only adds to that expense.

Paragraph 2 Summary

Based on the cited statistics about childcare expenses, reducing these would be the greatest benefit of a lengthened school day.

[P3] An additional advantage of an extended school day would be to allow for greater diversity and depth in curricula. Schools across the country have increasingly cut instruction in arts, music, and physical education (as well as recess) in order to meet objectives in reading and math. While this unfortunate state of affairs can be partially blamed on overzealous attention to standardized tests, it points to the larger deleterious trend of narrowing students' instruction. With a longer school day, such eliminated subjects can be restored, enriching students with a more well-rounded education.

Paragraph 3 Summary

Another benefit of a longer school day is enabling schools to offer a broader range of subjects and programs. In the author's view, this change would enrich students and improve their overall education.

[P4] [1] To this proposal, however, critics may object that the added time would impose strain on educators. Can we truly ask schoolteachers—already among the most overworked individuals in society—to endure even more hours in the classroom? The answer is that a lengthened school day need not distress teachers nor add to their already cumbersome workload. By providing for additional areas of study in the arts and humanities, the extension would give schools cause to hire new, perhaps specialized, faculty to offer these courses.

[1] **Addressing Opposing View**

The author raises an objection in order to address it: the worry that lengthening the school day would overburden teachers.

Paragraph 4 Summary

Lengthening the school day would not burden educators as some critics suggest, because additional teachers would be hired and some of

Moreover, the time could also be allocated to sports, academic clubs, and other extracurricular activities.

[P5] ² However, this point speaks to another objection, namely, the cost of adjusting the school day. Whether through paying current teachers more or hiring new ones, implementing such a proposal would entail a significant financial expenditure. There are at least two responses to this line of thinking. First is that this increase in the cost of schooling would be offset and likely surpassed by the aforementioned savings in childcare. Thus, while it is true that schools would require greater funding (likely necessitating higher property taxes), parents would ultimately pay the same or less overall, with greater educational opportunities for their children and fewer transportational burdens. Second is that schools should be better funded regardless. Recently, some schools—especially those in rural areas—have even reduced school weeks to only four days as a cost-saving measure. It is beyond dispute that schools across the board both need and deserve a radically increased investment from citizens. Lengthening the school day is simply one manifestation of how such funding should be utilized.

[P6] With this one change, states can coordinate the lives of parents and children, reduce the need for costly childcare, and expand curricular offerings. These worthy and desirable aims provide a clear justification for extending the school day.

Lengthening the School Day ©UWorld

---

the added time could go to extracurricular activities.

² Addressing Opposing View

A second objection is addressed: that lengthening the school day would be expensive.

Paragraph 5 Summary

The author agrees that lengthening the school day comes at a financial cost. However, they argue that these costs will be offset by savings in childcare and that schools should be better funded anyway.

Paragraph 6 Summary

Having explored the benefits of and potential objections to lengthening the school day (as promised in the introduction), the author concludes that the case has been made in favor of this proposal.

Passage K: Lengthening the School Day

Content Annotations

Passage Observations: 11 Annotations

[P1] There may be reasons to reject the idea of lengthening the school day. None of them, however, are *good* reasons. Rather, [1] the supposed demerits of such a proposal fall easily in the face of its numerous financial and social benefits for families.

[1] **Focus**

The purpose of the passage seems to be to defend the claim that school days should be made longer.

[P2] [2] The greatest of these benefits lies in reducing the need for childcare. It is a curious fact of American life that the adult's work schedule and the child's school schedule are misaligned. Children rise with the sun to head to classes, only to be sent home again hours before parents return from their jobs. In a society where, more often than not, both parents work, this discordance creates the need for an expensive arrangement to fill the gap in families' routines. [3] For instance, studies show that in 2016, childcare costs accounted for 9.5 to 17.5 percent of median family income, depending on the state. Today, 40 percent of families nationwide spend over 15 percent of their income on childcare. Transportation to and from care sites only adds to that expense.

[2] **Author's View**

According to the author, addressing childcare needs is the most important reason for lengthening the school day.

[3] **Outside Source**

The author cites data supporting the claims about the expense of childcare.

[P3] [4] An additional advantage of an extended school day would be to allow for greater diversity and depth in curricula. Schools across the country have increasingly cut instruction in arts, music, and physical education (as well as recess) in order to meet objectives in reading and math. [5] While this unfortunate state of affairs can be partially blamed on overzealous attention to standardized tests, it points to the larger deleterious trend of narrowing students' instruction. With a longer school day, such eliminated subjects can be restored, enriching students with a more well-rounded education.

[4] **Author's View**

The author's second reason for favoring a longer school day is increasing the variety of subjects that can be taught.

[5] **Author's View**

The author believes education should be broader and include more instruction in the arts and humanities, and that standardized tests have been overemphasized.

[P4] To this proposal, however, critics may object that the added time would impose strain on educators. [6] Can we truly ask schoolteachers—already among the most overworked individuals in society—to endure even more hours in the classroom? The answer is that a lengthened school day need not distress teachers nor add to their already cumbersome workload. By providing for additional areas of study in the arts and humanities, the extension would give schools cause to hire new, perhaps specialized, faculty to offer these courses.

[6] **Author's View**

The author seems to take for granted that teachers are overworked but argues that lengthening the school day would not add to this problem.

Moreover, the time could also be allocated to sports, academic clubs, and other extracurricular activities.

[P5] However, this point speaks to another objection, namely, the cost of adjusting the school day. Whether through paying current teachers more or hiring new ones, implementing such a proposal would entail a significant financial expenditure. [7] There are at least two responses to this line of thinking. [8] First is that this increase in the cost of schooling would be offset and likely surpassed by the aforementioned savings in childcare. Thus, while it is true that schools would require greater funding (likely necessitating higher property taxes), parents would ultimately pay the same or less overall, with greater educational opportunities for their children and fewer transportational burdens. [9] Second is that schools should be better funded regardless. Recently, some schools—especially those in rural areas—have even reduced school weeks to only four days as a cost-saving measure. [10] It is beyond dispute that schools across the board both need and deserve a radically increased investment from citizens. Lengthening the school day is simply one manifestation of how such funding should be utilized.

[P6] With this one change, states can [11] coordinate the lives of parents and children, reduce the need for costly childcare, and expand curricular offerings. These worthy and desirable aims provide a clear justification for extending the school day.

Lengthening the School Day ©UWorld

---

[7] Likely Question Topic

The phrase "at least two responses" suggests that what follows will be a list of reasons for the author's position. Such information could easily be the basis for questions about the author's view or the evidence for passage claims.

[8] Claim

The author argues that savings in childcare would help offset the added expense of a longer school day.

[9] Likely Question Topic

While the author's first response was about the effects of lengthening the school day, this second response is based on an independent claim that schools should receive more funding in general. This difference could be relevant to distinguishing a correct from an incorrect answer.

[10] Author's View

Claiming the point "is beyond dispute" suggests the author is taking it for granted and assuming the reader's agreement.

[11] Author's View

Here the author gives a concise summation of the benefits of lengthening the school day.

Passage L: When Defense Is Indefensible

This passage is associated with 5 questions analyzed in the UWorld MCAT CARS book.

- Subskill 2d. Connecting Claims With Evidence Question 2
- Subskill 3b. Passage Applications to New Context Question 3
- Subskill 3c. New Claim Support or Challenge Question 3
- Subskill 3f. Additional Conclusions From New Information Question 3
- Subskill 3g. Identifying Analogies Question 1

Passage L: When Defense Is Indefensible

Structural Annotations

Passage Structure: 8 Annotations

[P1] Suppose a prosecutor is considering whether to bring a case to trial. He is not sure that the suspect is guilty—in fact, based on the evidence, it's more likely that the suspect is *not* guilty. Nevertheless, he feels confident he can secure a guilty verdict. His powers of persuasion are considerable, and there's a good chance he could trick a jury into believing the evidence is strong instead of weak. In addition, the case is high profile and could be very lucrative; winning would likely lead to a substantial raise or promotion. He decides to charge the suspect, and ultimately succeeds in persuading the jury to convict.

Paragraph 1 Summary

In a hypothetical scenario, a prosecutor tricks a jury into convicting a man who is likely innocent so the prosecutor himself will personally gain.

[P2] Looking at this situation, most of us would easily judge the prosecutor as extremely unethical. His conduct is outrageous and wrong—he clearly acted with corrupt intent, perpetrating injustice in order to profit financially. Why is it not shocking, then, that we tolerate the mirror image of this behavior from defense attorneys? For they engage in the same outrageous conduct, only on the other side. Paid handsomely to represent even the vilest of clients, they apply their oratorical prowess to manipulating jury perception, keeping the guilty free and unpunished in exchange for money and status. To the extent that this behavior takes place, are some defense attorneys as unethical as our hypothetical prosecutor?

Paragraph 2 Summary

According to the author, some defense attorneys act just like the hypothetical prosecutor, which suggests they are just as unethical. So, the author questions why we don't seem to judge such attorneys in the same way.

[P3] It is worth distinguishing two senses of the word "ethical" here. For there are standards of *professional* ethics to which any attorney must conform, including standards particular to the defense. Most relevant to our purposes, a lawyer is obligated to provide their client with a "zealous defense." In other words, once an attorney takes on a client, they are ethically bound to promote that client's rights, interests, or innocence—in fact, *not* to do so would be *unethical*. Thus, one might try to suggest that this obligation undermines the claim that some defense attorneys act unethically.

Paragraph 3 Summary

The word "ethical" can mean two different things, and in particular there are *professional* ethics that defense attorneys must follow. However, the author hints that this fact still might not justify defense attorneys' behavior.

[P4] [1] However, meeting that professional standard is not the same as being ethical in the general sense of the word. The standard depends on the condition: *once an attorney takes on a client*. With the exception of court-appointed attorneys or public defenders, who are assigned to provide representation to those who would otherwise lack it, an attorney is never required to represent a defendant. Therefore, meeting one's obligations as a defense attorney does not necessarily make one ethical, because the choice to accept a specific case (and thus to incur those obligations in the first place) may itself be an unethical act.

[P5] Moreover, the role of court-appointed attorneys is to help protect the rights of citizens who cannot secure their own representation, usually for financial reasons. Although preserving those rights is necessary to uphold justice, this situation highlights how wealth and class contribute to *injustice*. While some defendants possess the means to hire top-level private lawyers, others must depend on public servants—frequently less experienced lawyers from overloaded, understaffed agencies. As a result, the rich are more likely to escape conviction even when they are guilty, and the poor are more likely to be convicted even when they are innocent.

[P6] It is doubtful that private defense attorneys could be somehow forbidden from representing guilty clients. Hence, the needed reforms to the system must come from individual attorneys committing to work for the right reasons. For those who strive to ensure citizens' rights, or who truly believe their clients are innocent, providing a defense is a noble undertaking. But for those whose overriding motivation is greed, [2] that legally "zealous defense" is ethically indefensible.

When Defense Is Indefensible ©UWorld

---

[1] **Connected Ideas**

The author now discusses the second half of the distinction made in the previous paragraph. There is a general sense of the word "ethical" that may differ from professional ethics, and upholding one is not the same as upholding the other.

Paragraph 4 Summary

The author distinguishes between public attorneys who cannot choose their clients and private attorneys who can. Since the private attorneys are not obligated to defend certain clients, their choice to do so may be unethical.

Paragraph 5 Summary

The author argues that the differences between private and public attorneys ensure that rich people are more likely to escape justice while poor people are more likely to be wrongly convicted.

[2] **Connected Ideas**

The author revisits defense attorneys' professional obligation to provide clients with a "zealous defense." As the author first suggested in Paragraph 4, private defense attorneys can still be unethical for choosing to take on this obligation in the first place.

Paragraph 6 Summary

Attorneys who truly believe their clients are innocent or who serve to ensure people's rights are ethical. But private defense attorneys who defend the guilty for money are still unethical. The system can be reformed only if such attorneys choose to give up that behavior.

Passage L: When Defense Is Indefensible

Content Annotations

Passage Observations: 9 Annotations

[P1] Suppose a prosecutor is considering whether to bring a case to trial. He is not sure that the suspect is guilty—in fact, based on the evidence, it's more likely that the suspect is *not* guilty. Nevertheless, he feels confident he can secure a guilty verdict. His powers of persuasion are considerable, and there's a good chance he could trick a jury into believing the evidence is strong instead of weak. In addition, the case is high profile and could be very lucrative; winning would likely lead to a substantial raise or promotion. He decides to charge the suspect, and ultimately succeeds in persuading the jury to convict.

[P2] [1] Looking at this situation, most of us would easily judge the prosecutor as extremely unethical. His conduct is outrageous and wrong—he clearly acted with corrupt intent, perpetrating injustice in order to profit financially. [2] Why is it not shocking, then, that we tolerate the mirror image of this behavior from defense attorneys? For they engage in the same outrageous conduct, only on the other side. Paid handsomely to represent even the vilest of clients, they apply their oratorical prowess to manipulating jury perception, keeping the guilty free and unpunished in exchange for money and status. To the extent that this behavior takes place, [3] are some defense attorneys as unethical as our hypothetical prosecutor?

[P3] [4] It is worth distinguishing two senses of the word "ethical" here. For there are standards of *professional* ethics to which any attorney must conform, including standards particular to the defense. Most relevant to our purposes, a lawyer is obligated to provide their client with a "zealous defense." In other words, once an attorney takes on a client, they are ethically bound to promote that client's rights, interests, or innocence—in fact, *not* to do so would be *unethical*. [5] Thus, one might try to suggest that this obligation undermines the claim that some defense attorneys act unethically.

[1] Author's View

The author believes there is wide agreement that the prosecutor's conduct is unethical.

[2] Author's View

The author argues that the behavior of many defense attorneys is just like that of the hypothetical prosecutor. Therefore, we ought to view these defense attorneys as being equally unethical.

[3] Focus

This question about defense attorneys seems to be the author's real point in talking about the hypothetical prosecutor. Thus, the passage's main idea likely concerns whether defense attorneys are unethical.

[4] Claim

The author distinguishes between two meanings of "ethical." In particular, the notion of *professional* ethics may affect how defense attorneys should be judged.

[5] Author's View

The author raises the possibility that defense attorneys' behavior might not be unethical after all because of their professional obligations. However, the phrase "one might try to suggest" indicates that the author may not find that position convincing.

[P4] However, meeting that professional standard is not the same as being ethical in the general sense of the word. The standard depends on the condition: *once an attorney takes on a client.* [6] With the exception of court-appointed attorneys or public defenders, who are assigned to provide representation to those who would otherwise lack it, an attorney is never required to represent a defendant. Therefore, [7] meeting one's obligations as a defense attorney does not necessarily make one ethical, because the choice to accept a specific case (and thus to incur those obligations in the first place) may itself be an unethical act.

[P5] Moreover, the role of court-appointed attorneys is to help protect the rights of citizens who cannot secure their own representation, usually for financial reasons. Although preserving those rights is necessary to uphold justice, [8] this situation highlights how wealth and class contribute to *injustice*. While some defendants possess the means to hire top-level private lawyers, others must depend on public servants— frequently less experienced lawyers from overloaded, understaffed agencies. As a result, the rich are more likely to escape conviction even when they are guilty, and the poor are more likely to be convicted even when they are innocent.

[P6] [9] It is doubtful that private defense attorneys could be somehow forbidden from representing guilty clients. Hence, the needed reforms to the system must come from individual attorneys committing to work for the right reasons. For those who strive to ensure citizens' rights, or who truly believe their clients are innocent, providing a defense is a noble undertaking. But for those whose overriding motivation is greed, that legally "zealous defense" is ethically indefensible.

When Defense Is Indefensible ©UWorld

---

[6] Claim

The author makes another distinction: public attorneys must represent whomever they are assigned to, while private attorneys can choose their clients.

[7] Author's View

Building on the distinction between public and private attorneys, the author argues that *the choice to represent* a client may be unethical regardless of lawyers' professional obligations.

[8] Claim

The public vs. private attorney distinction can also contribute to class-based injustice. The rich hire top-level private lawyers and so are more likely to go free, while the poor depend on less-experienced public attorneys and so are more likely to be convicted.

[9] Author's View

According to the author, there is probably no way to force private attorneys to represent only clients they truly believe are innocent. Therefore, the system can be reformed only if attorneys themselves decide to uphold that ethical standard.